Exposing Alaska

the Blessings and the Curses
Real Life Adventures

Don Elbert

Cover photo: The author's daughter Diane shows her delight on a snowy day.

Illustrations by Elaine Elbert, Denny Macom, Kurt Labonte, and Sean Young.

Final editing and layout: Sue Mitchell, Inkworks

Published by: Don Elbert

Copyright © 2002 by Don Elbert. All rights reserved. No part of this publication may be reproduced in any form or by any means without prior written permission.

ISBN 1-57833-176-5

Elmer E. Rasmuson Library Cataloging in Publication Data:

Elbert, Don.
Exposing Alaska : the blessings and the curses, real life adventures / by Don Elbert. – [Fairbanks] : Elbert, c2002.

p. : ill. ; cm.

Printed in Canada

1. Alaska–Anecdotes. 2. Alaska–Social life and customs–Anecdotes. 3. Hunting–Alaska–Anecdotes.
F904.6.E43 2002

Preface

Alaskans enjoy sharing their stories. Encounter some vibrant northern personalities as they reveal a reality more shocking than fiction. Join us in arctic comradery to learn why we think Alaska is special and why we live here.

These stories exhibit the struggles, joys, and fears of Alaskans as they tame a cold and inhospitable landscape. This collage of "mostly true" stories shows the humor, spirit, courage, ingenuity, successes, and sometimes failures of the people who live on the Last Frontier.

Acknowledgements

I wish to express my heartfelt thanks to my wife Elaine; sons Doug, David and Ken and daughter Diane; Kathryn De La Fuente; Alexandra M. Kienle; and Carolyn Petterson for their advice and help with editing stories.

My gratitude is also extended to Janet Baird and each member of the Fairbanks Community Writers Group for their assistance and guidance during our monthly workshops at Alaskaland.

This book is dedicated to the many Alaskans who shared their treasured memories and life-challenging moments.

Contents

Assaulted by the Big Ones
1   Attacked by Alaska's Giant Predators—*Don Elbert*

The Barrow Adventures
8   Journey to the Apex of Alaska—*Don Elbert*

Unforgettable Memoirs
22   Peggy's Dream Came True in Crooked Creek—*Don Elbert*

32   Mom's Tarpaper Shack—*Annette Bray*

35   The Fourth of July, Alaska Style—*Don Elbert*

42   Seven Moose on the Trail—*Cathy Webb*

44   I Held on to My Babies and Watched Our Belongings Float Downriver—*Ann Dolney*

47   Which One Jelly—*Ann Dolney*

49   A Lonely Birch Tree—*Don Elbert*

52   Dad and I—*Don Elbert*

My Faithful Dog Team
58   On a Cold and Snowy Night—*Richard Barnes*

Romance Blossoms in the Cold
68   Sometimes Snowbirds Fly North in Winter—*Cindy Caserta*

Growing up in Alaska
73   The Big Red Tractor—*Ken Elbert*

77   My Go-Cart—*Ken Elbert*

81   Experiences of a Twelve-Year-Old—*Dick Gilcrest*

83   My Worst Farm Experiences—*Dick Gilcrest*

Scariest Moments by the Pilots

85   Flying in a Whiteout—*W. T. Reeves*

94   We were in the Air but We Couldn't Fly—*Bill Griffin*

Bone Chilling Cold

97   Sometimes You Wake Up, Sometimes You Don't—*Bill Moberly*

105   I Don't Like Being Cold—*Bill Griffin*

107   I've Been in Seventy-eight Degrees Below Zero—*Bill Griffin*

109   Twenty-four Hours in Fifty Below Zero—*Tasha Custer*

111   The Indestructible Moose—*Paul Elbert*

Surviving Alaska's Relentless Waterways

114   Surviving the Chatanika—*Chad Dietz*

118   A Riverboat Nightmare—*Don Elbert*

135   I Dove into the Chitina River—*Mike Blackwelder*

Big Game Hunting in Alaska

137   Our Sheep Hunt—*Bill Griffin*

140   I Wanted a Moose—*Mark Neidhold*

149   My Best Moose Hunt—*Tom Keele*

153   No Caribou Steaks—*Paul Elbert*

156   From My Cabin Door—*Paul Elbert*

161   Hunting Bears—*Paul Elbert*

164   Youngest Hunter Kills Oldest Buffalo—*Douglas Elbert*

170   Triumphant in the Trophy Pasture—*Don Elbert*

181   Hunting Wolves—*Paul Elbert*

187   Wolves and Grizzly—*Paul Elbert*

190   We found a Den of Wolf's Pups—*Paul Elbert*

297   Antlers Coming—*Paul Elbert*

Assaulted by the Big Ones

Attacked by Alaska's Giant Predators

Don Elbert

Just minding my own business, I sat motionless, as old men do, on a park bench at the Chena Pump Campgrounds. My focus was on the Tanana River as its murky high water carried away uprooted green-leafed trees. A motor home with a Texas license plate grabbed my attention when it pulled up.

A lanky oldster swaggered out on bowed legs. He stood intently watching the river. Right away he began vigorously slapping at numerous mosquitoes that seemed strongly attracted to him. I noticed the dark tan burnt into his leathered face and hairless head. Then my eyes were drawn to his big-as-a-saucer belt buckle, which featured an embossed bucking horse. Moments later he looked my way and, between slaps, introduced himself. Then he started right out bragging, as they all do.

"Howdy, partner, my name's Charley Bowman," he says. "This muddy stream of yours here sure ain't nothing to look at compared to our cleaner and three times wider Rio Grande River down yonder in the great state of Texas. But I gotta tell ya, ya'll sure got some big hungry skeeters up here. Do they ever cause a body harm?"

Now, it is common knowledge that there is an ongoing rivalry of bragging rights between the residents of these two

∽ *Exposing Alaska* ∽

largest states. Seizing my opportunity, I replied, "Yup, they surely do, Charley Bowman. Come sit yourself down, and I'll tell you what they did to me."

While Charley made his way over, I recalled all the other loudmouth bragging Texans who had come up here for the big money during the pipeline days. And then they complained about freezing their buttocks off in the forty-below-zero cold. I got real tired of them stuffing fat paychecks in their pockets, all the while bad-mouthing Alaska, comparing our environment to a giant deep freezer. I quickly recognized my chance to scare the socks off at least one of those bragging Texans by telling him a tale. So as Charley slowly eased himself down, I began spinning my yarn.

It all started innocently enough, Charley; just a couple of guys heading out on a fishing trip with high hopes of catching supper. My skinny friend Bear Run, who was out of character that day by being sober, joined me, and we had driven up the Haul Road in my vintage pickup, heading into the Brooks Range. I parked, and with fly rods in hand, we walked the rocky trail, intent on finding a clear stream and hooking into several tasty arctic char.

Charley, I continued, I know about the dangers of living in Alaska. Flesh goes numb in minutes. Some folks have turned solid from exposure while still in the outhouse. I know moose, bears and wolves can attack when least provoked and stomp or chew you to death. But no one ever told me about them—Alaska's most unpopular species of wildlife: the Arctic's Giant Mosquitoes.

Fifty-five years spent in the safety of the interior of Alaska with normal mosquitoes had dulled my sense of danger. I'd spent weeks hunting moose out on the open tun-

∾ *Exposing Alaska* ∾

dra while substantial numbers from the twenty-six tribes of local mosquitoes hunted me. My previous success at surviving them had created my overgrown ego and left me with an I-can-handle-them attitude. However, nothing I experienced during those peaceful years had prepared me for the actual sighting of the Arctic's Blood Extractors.

Charley gasped, but I continued. My nightmare began a short thirty minutes after Bear Run and I had split up at a Y in the path. Wearing my bright red, lucky fishing shirt decorated with flies, I hiked on down into the valley, sloshing through the soggy tundra. Bear Run stayed on higher ground as he followed the trail around the far side of the mountain.

Little did I know that the enemy, those arctic mosquitoes, were waiting patiently until I got far enough into their territory so there wouldn't be any human witnesses to the massacre. The attack began with a sudden breeze that interrupted the calm of the early afternoon. Looking up, I detected what appeared to be a small helicopter preparing to descend upon me; but no . . . what I had mistaken for a rotor was the twelve-inch span of the wildly flapping wings on the dark watermelon-shaped body of a six-legged mosquito. She zeroed in on me with a parched look in her bulging yellow eyes, hovering above while scouting for a large vein with her five-inch erect stinger poised and ready to penetrate. If I had known what was coming, I would have turned and run; but in that frightening moment, I couldn't make my trembling legs move.

Mesmerized by the ponderous size of this arctic devil, I cringed, and that's when it happened. She skidded in, crash-landing on my sweaty forehead, knocking my eyeglasses off. She buried her size-sixteen syringe and my body

⌒ *Exposing Alaska* ⌒

exploded in pain. My eyes crossed and I fell on my knees on the verge of passing out as my assailant began siphoning blood. My head felt like I had just dived into the Chena Hot Spring's swimming pool—after they had drained it.

I sneaked a quick glance over at Charley. He was all ears and belt buckle, so I went on.

I'll call my assailant Miss Greedy. I nearly knocked my own self out with left and right uppercuts, punching and walloping my head in an all-out effort to crush her, but I missed every time and she just kept on slurping. An old-timer once told me that female mosquitoes just want your blood, but Miss Greedy was different. She wanted to hurt me bad. I saw vengeance glaring from her cruel, yellow golf-ball-sized eyes and tainted green bubbles of saliva dripping from her hideous snout. I was a sitting duck for what was coming—the rest of 'em!

Charley emitted another gasp.

Millions of ravenous mosquitoes flew in as if the cooks had rung the dinner bell at mosquito camp and I was the main course. Swarms of buzzing skeeters big enough to carry off sled dogs and small cars blocked out the sunlight and obscured the trail. Each one plunged her needle-sharp syringe into my weakened body and began feasting. During the brunt of the first assault I slapped, smacked, and crushed at least a thousand of 'em. They nailed me on every exposed patch of skin. The airborne invaders even flew up my nostrils with each inhalation, plugging the airway and forcing me to hold my breath until I nearly passed out. The overpowering need for oxygen caused me to open my mouth, and I instantly sucked in even more mosquitoes. Choking, gagging, sputtering, and spitting disoriented me and I swallowed the rest.

∾ *Exposing Alaska* ∾

Here an engrossed Charley asked, "Didn't you fight back or did you just give up?"

Oh no, Charley, I assured him, I didn't quit. I knew this was going to be a fight for survival, and I was flailing away at 'em. On my better slaps I nailed anywhere between fifteen and thirty foes at a time. My own blood spurted from their smashed bodies and now the tundra was stained red. But they just kept coming, from a seemingly inexhaustible supply. I really believe they were breeding faster than I could squash 'em. Mosquitoes came from all directions, honing in with needle sharp beaks, jabbing me through every thread of clothing. Each one tanked up and left to make room for the next in line.

I suddenly realized that if I were to survive them, I had to resort to desperate measures. I gathered my last ration of energy and broke away in a dead run. In my weakened condition, I had to stop every hundred yards to catch my breath. They caught up and the mugging would start all over again. By the time I had run half a mile my legs just quit moving, and I fell to the ground in a heap. I looked back and to my immense relief the mosquitoes were far behind. Starving for oxygen, I began gulping in air, when a deafening roar like a revved-up Harley motorcycle caused me to whirl around. That swarm I'd left behind must have cellphoned ahead because a larger collection was waiting in ambush. I had innocently stumbled into the heart of their domain, and they overwhelmed me. A zillion or more mosquitoes covered me like a black blanket.

I sneaked another peek over at Charley. I could see his tonsils as he stood open-mouthed, staring at me. He appeared to be in shock, so I stifled my gloating and went on.

5

Exposing Alaska

One week later I woke up in intensive care. The blood transfusions had returned a spark of life to my body but I still remained heavily bandaged. I braved a quick glance in the hospital mirror and nearly fainted when a ghastly swollen mummy stared back at me. I have woken up many nights since in a cold sweat, screaming, as I relive the near-fatal attack.

That afternoon, Bear Run stopped by and told me how he found me. "Not seeing any sign of a creek ahead," he explained, "I came trotting back looking for you. I noticed this large ball of mosquitoes in the middle of the trail and stopped to see what they were eating. I felt sorry for the poor victim. Then I noticed a flash of bright red and I recognized your shirt. I knew it was you that they were eating alive! In my desperate attempt to rescue you, I waded in with both arms swinging and must have squashed a wheelbarrow full of 'em. Then I pulled your limp body from the remaining hordes and flung you over my right shoulder. With my own strength waning, I rushed you back to your pickup and floor-boarded it all the way to the emergency entrance of the Memorial Hospital."

My doctor stopped by later and told me, "When they first wheeled you into the emergency room, I looked you over and discovered you had no pulse and in fact your veins lay flat. You didn't have a drop of blood left in you. I was pulling the white sheet over your head when Bear Run screamed."

"No Doc! You gotta try to save him!"

Doc replied, "I really thought it was too late but I hooked you up to a gallon of blood from the blood bank and let it seep in over night. It is a real miracle that you pulled through."

∽ *Exposing Alaska* ∽

I've been out of the hospital one month now, Charley, and I'm slowly returning to my old cantankerous self. This is the first day I've strayed far from the confines of my cabin. A deep-seated fear that those Arctic Mosquitoes followed me home has forced me to delay trips to the outhouse for two or three days at a time.

I glanced over at Charley. His eyes had grown as large as his belt buckle, and he stared at me with a stone face. Several minutes passed before he spoke, "Wow! We got poisonous rattlers in Texas but they ain't nothing compared to those Arctic mosquitoes that ravaged ya'll. Ya'll can have your big old Alaska, your muddy Tanana, and them giant mosquitoes. I can't figure the attraction that keeps you northerners up here in harm's way, unless all those mosquito bites have seriously affected your brain. I'm heading back home and if you had a lick of sense left in ya, you'd consider moving to a safer place like Texas!"

I had to stifle my elation at the effect my story had on Charley while I observed the numerous mosquitoes that were practicing their touch-and-go landings on his bald head. Charley slapped himself again in self-defense then turned and scurried on shaky legs toward his motor home. He sputtered several derogatory remarks about those bothersome mosquitoes as he bounded up the steps and slammed the door. He roared off before I had a chance to wish him a safe journey back to the Lone Star State.

The Barrow Adventures

Journey to the Apex of Alaska

Don Elbert

Our first jaunt to Barrow started on June 22, 1997: my outdoorsy wife Elaine's birthday. She yearned to see Alaska's most northern village, a settlement of 5,000. We had read and heard tales about this intriguing place. So hand in hand, we boarded an Alaska Airlines jet to experience life on top of Alaska.

We hiked to the Arctic Ocean and observed huge hunks of blue-tinted ice bouncing like corks in the choppy waters. We watching the local kids ice-hopping from one floating chunk of ice to the next floating ice cube under twenty-four continuous hours of golden sun. This appeared to be great fun for the fearless residents, but Elaine or I were neither anxious nor brave enough to risk a possible cold dip in the ocean.

Then we walked under a fifteen-foot-tall whale jawbone arch on the beach. I wondered about the size of the whale's mouth that supported those jowls. I figured back when it was alive, this whale could have opened up and swallowed a deluxe-sized car.

Two months after returning home, our eldest son David accepted a teaching position as the electrical instructor at the Ilisagvik College in Barrow. (Ilisagvik means "a place to learn" in the native Inupiat language). David bore a look

8

of distinction befitting an instructor. He had a caring attitude, a calming voice, and a touch of gray at his temples. So, in September of 1997 he began teaching the first electrician course offered in the North Slope Borough.

David had to leave his family in Fairbanks the first year to fulfill this commitment. He spent his free time checking out the scarceness of family housing. Eventually he found a fixer upper, moved in, and began remodeling. The second year his charming wife Kelly and four of their five talented children joined him in Barrow for the 1998 school year. Jaime, their oldest child, had graduated from high school and currently works in Fairbanks.

In May 1999, David's son Michael graduated from Barrow High School. The occasion called for our return to Barrow for the ceremony and festivities.

"What? Barrow? Whadda ya flying up there for?" asked my grizzled old neighbor, Sam. "Going to Barrow is like moving from your living room into your freezer. Anyone with a lick of sense would be heading south. You're sufferin' from frostbite on the brain, Don!"

Sam was referring to the fact that the Inupiat Eskimo town of Barrow is on the northernmost tip of Alaska, 506 miles north of Fairbanks. Perched on the edge of the frozen ocean, it is still locked in cold long after spring has arrived below the Arctic Circle.

Our family departed Fairbanks on May 13. Sam had taken us to the airport and now stood shaking his gray-whiskered head as we boarded. Elaine and I, along with our charming eighteen-year-old granddaughter Jaime, and our daughter-in-law's parents, Dick and Barbara, waved goodbye to Sam and the Golden Heart City. The Alaska

~ Exposing Alaska ~

Airlines Boeing 737 lifted off the runway into clear blue skies and we were on our way to Barrow to participate in our grandson Michael's high-school graduation.

A crisp, cold blast of wind greeted us as we walked to the small airport terminal in Barrow, but once inside, we were warmed by hugs from our extended family. After collecting our luggage we all scrunched into David's van and drove the mile to his unpainted plywood house mounted on pilings a foot above the ground.

That evening, David drove us to the Barrow High School. A large sign on the side of the school announced that Barrow is the "Home of the Whalers." Whalers are the bold and brave Natives who disregard danger to hunt huge whales from lightweight walrus or sealskin boats in icy waters. It is an integral part of life at the top of the world.

Several minutes after we arrived, our grandson Michael Elbert, wearing a broad smile and a blue gown over his boxer shorts and tennis shoes, strutted triumphantly into the Barrow High School gym with thirty-eight classmates. Each happy graduate proudly marched under an arch of baleen held by students from the junior class. Every step timed to the rhythmic beat of sealskin drums. The program opened with the Pledge of Allegiance, recited in both English and Inupiat, and then the Barrow Dancers performed their native dances.

Through misty eyes of pride I watched a radiant Michael walk across the stage in front of his dignified teachers to receive his diploma and honors. Michael had blossomed into a tall, handsome, marvelous young man.

On the morning of our second day in Barrow, Michael took us on a drive around town. That didn't take long. He pointed out that there is only one bank in Barrow and one stoplight, which mostly flashes a cautionary yellow. There are two grocery stores; both were selling milk for seven dollars a gallon. Supplying other needs to the locals were four video rental places, a hospital, one gas station, one computer store, and one company with a monopoly on Cable and Internet services.

The largest of the several restaurants in this farthest north city and Michael's favorite is Pepe's. This out-of-its-element eating place is the northernmost Mexican restaurant in America and probably the world. After our noon tacos we all agreed with Michael's choice. Pepe's serves excellent Mexican food.

✑ *Exposing Alaska* ✑

The lone gas station in Barrow receives its supply of gas once a year during a brief thaw in the otherwise frozen Arctic Ocean. The laden barge slips in, unloads its cargo of gas and supplies, and departs before the ocean refreezes. The town's one station sells gas all year long for $2.61 a gallon. Comparatively, the day we left Fairbanks the sign at our local Tesoro station displayed $1.26 for a gallon of unleaded.

The streets of Barrow are not paved, and in the four days we were there, the road surface began to change from winter hardness to softer summer slush. Mud splattered the snow that skirted both sides of the road and we left interesting footprints as we walked along the streets.

Michael told us all the fun things to do in Barrow: spotting whales in the ice leads, searching for polar bears or seals out on the ice, having snowball fights, riding your snowmachine or four-wheeler, climbing on icebergs and ice-hopping. He went on to explain that when the frozen Arctic Ocean breaks up, some kids and a few not-so-grown-up adults attempt to see how far out on the ice they can get, hopping from one floating cake of ice to the next. This is not recommended recreation because people do fall in; and yes, it is ice cold!

Houses are built on buried pilings in Barrow because the ground is permanently frozen. Many houses are small, especially the older ones, and many are unfinished and unpainted, like the outside of David's. Most homes are heated with natural gas. The landscape around Barrow is a white expanse of flatness and emptiness and the wind blows much of the time.

For eighty-three days from May 11 to August 1, the sun never drops below the horizon. Not that it makes much difference, since most of the heat radiating from the faraway sun is dissipated before reaching the earth at Barrow. The summer temperatures vary, but average around forty degrees Fahrenheit.

Michael told us of sights we would not see in Barrow, "penguins, four lanes of traffic, bumper-to-bumper cars, anyone aside from the local police officers wearing a seat belt, or any roads leading far from town. You won't see any trees or mountains for a hundred miles from Barrow, and you won't find a movie theater, a McDonald's or even a shopping mall. You won't see anyone out sunbathing, and in fact, after the sun sets on November 18, you won't see it again for the next sixty-five days in a row."

Our northernmost Elbert family had arranged a Barrow treat for all of us that afternoon: a two-hour dog sled ride out on the Arctic Ocean ice. Their friend and the owner of Arctic Mushing Tours, John Tidwell, Jr., let us harness up

Standing on top of an ice mountain on the ice of the Arctic Ocean near Barrow.

the team while he spent his time picking up the numerous piles of dog poop.

We helped load his sleek wooden dog sled into the back of his pickup. His ten-dog team plus one extra dog were loaded in their own special trailer. We drove through the muddy streets to the edge of the frozen ocean. There we unloaded and hooked ten of the barking, can't-wait-to-run huskies to the sled. John uses two lead dogs instead of the usual one. I asked John, "Why the eleventh dog? Why do you bring along a spare husky?"

He replied, "Oh, the extra dog just runs alongside the team, but if we encounter a polar bear, that one dog will instinctively charge and attack the bear. That gives me time to race the team and my passenger back to safety. I would rather lose one dog than all of us in a polar bear encounter."

We brought along four snowmobiles to accommodate the eleven people in our party. Each of us took turns mushing the dog team, riding in the sled or driving the snowmobile.

After thirty minutes of enjoyable riding we stopped one mile out from shore and climbed on top of a thirty-foot-high ice heave. It looked like a giant ice cube pushed up on top of the ocean's ice. Our grandchildren, overflowing with

Elaine Elbert at Point Barrow.

A thirty-foot-high ice heave pushed up on top of the frozen Arctic Ocean.

energy and agility, led the way. I really felt like I was on top of the world, as if I had reached the summit of Mt. McKinley, the largest mountain in North America. Everywhere I looked was down. Elaine was holding my hand as we took in the spectacular view together. I told her, "Sure wish Sam was here to enjoy this. Then he would understand why we came."

Three-quarters of a mile north, we could see a wide lead, or break in the ice, where whales are sometimes sighted. Searching with binoculars, we did not see any sign of whales, but we did spot a seal sunbathing on the edge of the lead. Watching closely, John, our experienced guide, pointed to a polar bear slinking up on the ice, edging toward the seal. John explained, "Polar bears are hard to spot out on the snow-covered ice because they're all white. They are smart too. They cover their black nose with a paw while creeping up on a seal. You have to look for movement to detect them." We watched the bear closing the gap when suddenly the seal raised his head and slithered from the ice. That polar bear didn't have that seal for his noon lunch that day. The following day, John Tidwell reported that three polar bears had been spotted just off shore.

Our four days of thrills and excitement in Barrow came to an end, and our graduate grandson Michael returned to Fairbanks with us.

Midway through the flight, Elaine whispered in my ear, "I want to return to Barrow for another visit when the ice goes out and watch them bring in a whale."

I smiled, "Let's just go back next year and take a chance on seeing a whale." Then my thoughts focused on returning home.

After touching down in Fairbanks and rolling up to the gate, we walked inside our big modern airport. I spotted Sam there to meet us. His eyes glared out from his bearded face, which hinted of an I-told-you-so attitude. Sam was cagey though. He didn't utter a sound. He quietly waited for a response from one of us.

I just blurted out, "Oh, Sam, you should have come with us. We had a blast! What a trip! Michael's graduation and all kinds of family fun in a really different place! The warm glow radiating from family love kept all of us from freezing. That's why we went, Sam. That's the reason we were willing to step back into winter, and it turned out to be so special for each of us."

Sam's hard features broke as he smiled his understanding. Then Sam said, "Well, when you put it like that . . . I'm right proud you had a grand time!"

By mid-September of 2000, plans for another visit north were shaping up. This time a whale sighting dominated the time for our departure to the top of the world. Plane tickets were purchased for September 29, in anticipation of arriving for the opening of the whaling season before the ocean froze. Elaine kept her eyes on the newspaper and her ears open for any news of whale activity in Barrow. The next morning she read in the newspaper that the Native whalers had delayed the hunting season a week. She grabbed our phone and called Alaska Airlines to delay our departure date one week. Three days later, we received a call from David telling us that the natives had just brought in a whale. We felt let down but still held hopes for the success of another Barrow whaling crew repeating this feat while we were there.

∾ *Exposing Alaska* ∾

Elaine and I departed Fairbanks in the cool twenty-nine degree temperatures at 5:00 PM Friday, October 6. Our plane landed smoothly in a slight wind at 6:40 PM in the twenty-one-degree chill of Barrow. David picked us up at the airport with great news. Two boats had radioed that they were each bringing in a whale at 7:00 PM!

The ground in Barrow was coated with a two-inch blanket of snow, similar to Fairbanks. We drove the four miles east and approached a large open beach area in the fading light of day. There were numerous cars and trucks on the beach. Many of the vehicles had gathered around in a hundred-foot circle with their headlights on and engines running. David pulled in between two trucks, and we had a ringside seat for the giant attraction: the butchering of a whale. However, we didn't stay in Dave's car long. We jumped out into the coolness of the night and joined the small crowd milling around taking pictures and observing this enormous carving ritual.

My first impression of what lay on its back in the center of the circle was a beached submarine. As I walked around this giant sea monster I felt so small and also so pleased that Elaine and I were finally going to see a whale being harvested. It was a bowhead whale, with baleen for straining its ocean food. The baleen sticking out of this whale mouth reminded me of a four-foot-long oversized comb. A lot of activity was happening on and around this black giant. Captain Savok and members of his crew had begun the carving. As the daylight faded into darkness, generators and floodlights were brought closer to illuminate the area enabling the night's work to continue. U.S. Fish and Wildlife agents were gathering data, taking blood specimens and measuring the length. "This whale is a young,

18

small female," the agent stated, "just thirty-two feet, seven inches in length."

"This young, small female is longer than my motor home," I told him.

A crewmember, using a huge, sharp long-bladed flensing knife mounted on a four-foot handle, cut through the three-quarter-inch hide up near the head of the six-foot-high torso straight down to the ground. He continued making cuts every ten to twelve inches all the way back to the tail. Other members of the team with additional flensing knives worked to deepen those cuts through the eight inches of blubber. Then one of the more agile men climbed up on top of the whale with a large ulu knife mounted on a longer pole. He made the vertical cut and began peeling the first strip of blubber back. Three or four others would sink their three prong hooks into the blubber strip and begin pulling while the man on top kept cutting the blubber free. When they reached the bottom of the whale, they made a final cut and slid a nine-foot long strip of blubber away from the carcass. Others would cut these long strips into eighteen-inch lengths and sort them into piles. The first

and largest portion went to the captain, and each of the eight-man crew had his own pile.

Once all the blubber was peeled off, the crew started on the meat which was now exposed and such a dark red that it looked black. This meat, also a delicacy, was cut off in strips similar to the blubber and added to each whaler's growing goodie pile.

The driver of a four-wheel-drive pickup pulled the rib bones away from the innards. We didn't notice any unpleasant odor because they were careful not to puncture the intestines, at least not while we watched.

One hundred and fifty feet away the other crew and their families were performing the same procedure on their whale. This was a thirty-foot-long male brought in by Captain Patkotak and his crew. Captain Patkotak, still beaming with pride from his successful hunt, invited us to taste and enjoy their choice muktuk.

We observed the gleeful faces and listened to the smacking lips of all that were chewing on this sought-after delicacy. The first small bite was more than I needed. "Chawing on muktuk gives a whole new meaning to chewing the fat," I emitted.

Elaine added, "It's kind of bland and greasy, and the longer I chew it the bigger it gets." We continued chewing and walking back and forth to see as much as we could until 11:00 PM.

In the morning, we returned to the previous site of high adventure to see what was left: some of the guts, a pile of bones, and a big circle of blood saturated snow. Eskimos don't waste any part of the whale.

Several hundred sea gulls were squawking with joy while they feasted on everything edible. We heard that the butchering went on for seven hours, until 2:00 AM. David told us that any of the remains would be moved further out to the point for the polar bears. This is an ongoing effort to keep the bears away from town.

Sunday afternoon before departing, Elaine confided in us all, "All my Barrow dreams have come true. Seeing our family happy and coping with village life, topped off by finally seeing a whale being harvested has made the last few days so rewarding."

Elaine had a look of pure contentment on her pretty face and a whale of a smile as she hugged and said her good-byes to family. She wiped tears of joy from her eyes while walking contentedly toward the Alaska Airlines jet.

Muktuk, the skin and blubber of a whale, is an Eskimo delicacy. It is eaten fresh, frozen, boiled, or fermented.

Polar bears are symbolic of the Arctic. These great white bears are a mixture of beauty and power. They live on drifting ice and swim in icy water. Their main prey is seals. A polar bear in good shape needs about two hundred pounds of seal meat a day. These bears can run thirty-five mph and are the only bears that have hair on the soles of their feet. It is common for a mother polar bear to deliver two tiny cubs weighing 1 to 1.5 pounds each. Due to mother's rich milk, these cubs can weigh twenty-five pounds by the time they leave the den in the spring.

Unforgettable Memoirs

Peggy's Dream Came True in Crooked Creek

Don Elbert

The lure to the vocation of enlightening young children can be a strong attraction for a young woman from Dillon, Montana, and that's admirable. But when that woman dreams that unfulfilled vision for most of her sixty years and then undergoes the training to become a teacher, now that is downright commendable.

A few months shy of her sixtieth birthday, the age that many women tend to shift into slow gear and ease into the soft lane of life, Peggy Birkenbuel did the opposite. This short, stout, talkative woman with a ruddy complexion and graying hair went back to school. She proceeded to fill her head with up-to-date knowledge and teaching skills. With her eyes wide open and a heart overflowing with love, this spunky woman overcame the numerous challenges to a senior citizen entering the teaching profession. She completed her college training and earned the credentials to be an elementary school teacher.

Peggy spent the following three years substitute teaching. She soon felt confident and up to the challenges of a full-time teaching position. However, after filing more than seventy applications throughout rural and urban Montana without success, she determined it was time to expand her search. Peggy had just recently read the novel *Tisha* and

∾ *Exposing Alaska* ∾

felt akin to this young Alaska teacher. A desire emerged from that reading which led her to seek a job in Alaska.

Using her meager savings, she ventured north in April of 1997 to attend the Anchorage Teacher Fair at the Captain Cook Hotel. Her hopes were high in the beginning, but her spirits soon sagged as her chances for a job diminished. The younger teachers were the ones receiving the contracts. However, during the four days of interviews, Peggy did earn the "Most Spirited Teacher Award" from among the 1,200 attendees. The prize was a hooded gray sweatshirt donated by St. Mary's village school.

Although she had numerous interviews, she only received two promising offers. The first offer fell through when that school burned down. The next job pitted her against a younger male teacher possessing the same credentials plus an endorsement in special education. He won out. Internally Peggy mused, "What was lacking in credentials I felt would be made up for by my years of life experiences and survival skills."

Crestfallen, she returned home to Dillon without a contract. She picked up where she left off, occasionally substitute teaching, housecleaning, and working with the elderly in home health care through the local hospital. This seemed to be her fate, since it had been five months since the teacher fair. Any hope for a teaching position that late in the year seemed slim.

Then, on the last Monday in September, the phone rang at 3:00 PM. The upriver principal from the school district office in Aniak, Alaska, invited Peggy to an on-line telephone conference interview. She accepted and appropriately answered all the questions. Peggy felt encouraged when the principal promised to call back mid-week. The

23

principal had set up connections to Crooked Creek with the village president and school board representative plus a parent for that conference call.

Wednesday afternoon her phone rang only once before she snatched it up with confidence. The first questions from the committee centered on the living conditions and were more like testing the depth of the waters to see what might be cause for her to not accept their offer.

"The house has a very low ceiling," the principal pointed out.

"I'm only five feet two inches," she replied.

"It's an awfully long walk to the school from your house," he said.

"I enjoy walking; it's no problem." Her confidence increased.

"It gets terribly cold here," he cautioned.

"The temperature drops to forty below in Montana regularly," she shot back.

"The winter months are long and darkness lasts for four to five months. This remote village is only accessible by boat or barge in summer or plane depending on the weather conditions," he warned.

"There will be plenty of school work to keep me busy enough," she countered.

The interview lasted over an hour. Peggy had a ready answer for every challenge the hiring committee posed. In her heart she glowed with confidence, and yet she felt the twang of uncertainty. She knew this could be her one and only chance to get a job in Alaska.

Reaching for her Alaska map, Peggy located the village of Crooked Creek on the banks of the Kuskokwim River. She thought, well, nothing there could be more difficult than the adversities she had already encountered in life as a single mom, raising three energetic boys. She had learned her winter survival skills sufficiently while enduring the cold Montana winters, as a National Ski Patrolwoman and ski instructor over the past thirty-five years. Peggy loved the outdoors and had confidence she would fit right in with the village life in bush Alaska.

Her last instructions were to wait until Friday, October 1, and be near the phone at noon to get the board's decision. At 1:00 PM Friday the phone rang as the promising applicant puttered in her kitchen, whiling away the time baking cookies. The connection faltered, but the definite message coming from the upriver principal was an emphatic, "You're hired!" Then he added, "You'll have to be here by Oct. 4, before school starts, if you want the job."

This offer, which resulted in her first full-time teaching job, was at the far away Johnnie John, Sr. Elementary School in the isolated Yup'ik Eskimo village of Crooked Creek in the southwest corner of Alaska. With a happy heart she accepted with a very vocal, "YES!"

Peggy promised she would report in on Monday morning the fourth, just three days away. The good news barely had time to sink in before travel plans filled her head. Arrangements had to be made quickly. No time to waste gloating over the joy of being accepted. Purchasing airline tickets, packing, and telling her sons that she was leaving to teach in bush Alaska was her agenda.

Armed with her life long dream and new teaching skills, Peggy was thrilled with the excitement of moving to a remote village. This petite woman with an abundance of energy and a ready smile packed a few belongings to carry and a dozen boxes, which she shipped ahead to furnish a kitchen and bathroom.

25

∾ Exposing Alaska ∾

She stopped at the 7-11 store where her sons Cory and Bret worked. She visited the meat market where son Brady was cutting up game meat from the past hunting season.

It took a near miracle of airline scheduling for her to be in Cripple Creek in three days. Finally, the two-seat Yute Airlines plane touched down on the hard-packed dirt landing strip in Crooked Creek on October 4, 1997. Earl, the pilot, on hearing her stomach growling and knowing the trading post would be closed, gave her his lunch, two cheese sandwiches, before he departed.

Peggy took a long first look at this Yup'ik and Athabascan village, which is carved out of the spindly spruce trees attempting to grow in the damp tundra along the Kuskokwim River. She liked the wilderness setting right away. In her heart she felt that this was the perfect spot for her to concentrate totally on teaching. However, no one was there to greet her.

Ten minutes later a scraggly fellow with a flat, runny nose dropped by in his rusty pickup with the tail pipe dragging.

He asked, "You—new school teach-eer?"

Peggy nodded, "Yes."

The man gave her a sly smile before introducing himself, "I'm Freddie I'll give you a ride to your living quarters if you like."

Peggy accepted and listened to the tail pipe rattling on the rutted road as they traveled. Upon their arrival at a small shack, Freddie said "That it," and just dropped her off.

She frowned as her eyes took in the weather-beaten shed. A glance inside brought a scowl. Later, she learned that the locals called it "the cave." Mary, an elderly neighbor, stopped by and, in broken English, welcomed her to the village and to her meager living conditions. Her

26

*~ **Exposing Alaska** ~*

two-room living quarters contained a bed, sofa, a camper-size refrigerator, coffee table, a metal folding chair, and a small bathroom.

The town's only store was the trading post located in the basement of a person's home. It was closed and wouldn't open again until the following Monday. She didn't have any food, her clothes hadn't arrived, and there were no restaurants. There were a few cleaning supplies and a pair of yellow rubber gloves, so she cleaned her toilet. (It had about an inch of frothy orange rust floating on top of the water.)

Peggy was the third teacher hired by the school district to teach the rambunctious second- and third-grade students. The first teacher they had hired only lasted three days. The second teacher endured eight days before calling it quits. She was told that the primitive living conditions had got to them. Peggy was used to hardships, being a single mom and raising three sons. She often worked three jobs to make ends meet. Teaching had been her dream, and she wasn't about to let these inadequate living conditions dampen her spirits.

Even though Peggy thought she was prepared for the meager accommodations that she had heard about living in Alaska's wilderness—she wasn't. Arriving with only the sympathetic bush pilot's donated lunch, she spent the first few days wondering what she had gotten herself into. In retrospect, she wished someone had impressed upon her the lack of necessities in the bush. However, her first priority was in finding a stick big enough to keep the mean loose dogs away.

Her stomach went into shock from deprivation the first two months and she lost 30 pounds. There were many culture differences, but she learned how to survive in this

27

village of 120 natives, where she was known as one of the few gussuks, or white persons, in town.

Peggy's hopes and dreams came true in Crooked Creek and parents' comments like this one fed her soul: "For the first time, I now see a glow in my child's eyes when she talks about school. My little girl loves going to school, and she loves Ms. Birkenbukel." That's the kind of love and inspiration that keeps Peggy teaching. She had longed to be a teacher since childhood, and now as a senior citizen she had reached her goal.

A news story printed in her hometown paper back in Dillon, Montana, about Peggy teaching in bush Alaska resulted in her answering over two hundred personal letters all winter to her inquiring and well-wishing new pen pals. The following letter is one of Peggy's replies:

Dear Pen Pal,

Thanks for responding to the news story. In addition to teaching all day in the class room and my correspondences with two hundred-plus pen pals, I am taking three credits of Alaska history class via audio conference on Friday from 5:00 PM to 7:30 PM and a two-credit reading endorsement course by correspondence. I study well into the night but love every minute of it.

Now I'll answer a few questions generally. No, I do not have television at my cabin, no tape player, and no washer-dryer either, but the school has a washing machine. There are thirty-three students in Johnny John Sr. School, taught by four teachers. One teacher has kindergarten and first grades, and I have fifteen students in my second and third grade class. The fourth through eighth grades have one

28

teacher and high school has the other. We all have small classrooms.

For you to really get a feel of what our village is like here at Crooked Creek, read some Alaska history books. Our village is 275 miles west of Anchorage and is just like it was in the early 1900s. The land is the same as it was from Russian settlement and early exploration. The second largest river in Alaska, the Kuskokwim, swings in a huge horseshoe out front. We are a fishing village and have been here for over a century. I have noted exploration, dating back to 1844 by Russian Naval ships coming upriver; next came the fur trappers and then the gold rush prospectors. A few gold claims date back to 1869. Crooked Creek flows into the Kuskokwim River near my cabin.

The hunting and fishing are extraordinary. I've watched wolves chasing caribou out on the river. The bears will be coming out of hibernation soon.

All winter long, I love walking the mile and a half to school at 7:45 in the morning in the dark. And then walking back home at 7:00 that night in the dark again over the frozen river. The cold and the darkness didn't bother me. I dress warmly and the walk keeps me in shape. These walks are special moments of peacefulness for me. The starlit nights are spectacular. When the moon is shining brightly, it adds a glisten that reflects off the white snow and gives light to my way.

Planes come delivering mail and an occasional passenger when the weather permits. The airstrip is a quarter mile from the school. We spend ten-hour days at school plus many weekends. We can use the washing machine to wash our clothes but have to take them home to dry and they often freeze on the way. A wire stretched from one side of my shed to the other supports my clothes for drying. I continually run into wet clothes. Teachers can take a bath in the schools single bathroom after classes.

There are no jobs in the village except schoolteachers, cook, janitor, bus driver, two aides, post office, health aide, and electric/oil man. Residents live off their Alaska Dividend, government checks, and food stamps. The children are the ones who suffer. I hope and pray that just maybe I can make a positive difference in their young lives. They have named me in the Yup'ik language Abyuk, meaning bright star. I have eaten agutaq—Eskimo ice cream, which is boiled fish mixed with Crisco and frozen blueberries. I much prefer plain old strawberry ice cream. Currently, the river ice is breaking up and worries of flooding persist.

The trading post gets groceries from the mail plane but not many veggies, mostly apples, oranges, and black bananas. I have adjusted to life in the bush without running water. What I really miss and look forward to the most is a long and soothing hot bath. Just two undisturbed hours soaking in a very warm tub of comforting sudsy water. Next, I would like to wash the mosquito spray and sand out of my hair and to go sleep on clean sheets. My shrunken and deprived stomach has been screaming for a restaurant meal, topped off with a double-

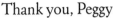

decker strawberry ice-cream cone. My house now has a couch and a coffee table, and I'm developing trust and friendship with the locals. Yes, I do miss my three sons, but we write frequently.

Thank you, Peggy

Another popular recipe for Eskimo ice cream is a whipped mixture of berries, seal oil and snow. It has its own distinct flavor, and is still not available at Baskin Robbins.

Originally, mukluks were boots made of the skins of a mukluk. "Mukluk" is the Yup'ik Eskimo word for the bearded seal.

31

Mom's Tarpaper Shack

Annette Bray

My husband Don and I, with our five young children in tow, moved from Delta Junction to Fairbanks in the fall of 1971. The ages of our children ranged from our newborn and first girl Susan, up through Paul, age seven. Don is an expert carpenter and he planned to build a home for us on our recently purchased land in Goldstream Valley, before the cold winter set in. In the meantime the seven of us were packed tight in our ten-and-a-half-foot camper, mounted on the back of Don's pickup. Then Don got a call from the union to help build a school in Anderson, eighty-five miles to the south, before we got the foundation laid out. I did not want to spend the winter crammed like sardines in our little camper, so Don quickly helped me draw up plans for a temporary shelter, a sixteen-by-twenty-four-foot house that came to be known as Mom's tarpaper shack.

Don returned home on weekends and laid out our work for the week ahead. That way, the kids and I could build our new structure. We began as my husband told us, laying two-by-eight-inch floor joists upright on the ground; next we added insulation and then nailed down the plywood, which became the floor. Once the floor was in place our goal was to complete one wall each week. The drop in temperature each night was a constant reminder to keep

∽ *Exposing Alaska* ∽

working because winter was fast approaching. However, my time was split between taking care of my babies, cooking for and mothering my boys, and being head carpenter. And I'm not even a good apprentice; I was just lucky enough to marry a carpenter, and oh, I wished he was with us now.

It soon became very evident that I had to keep our three older boys busy. So I took the time out to buy three small hammers. Then I would measure and cut each two-by-four, lay them out on our new floor and nail the pieces together, but I only drove the nails in half way. Paul, Jeff, and Gene finished driving the nails home. They were helping and becoming good carpenters, following in their dad's footsteps. Each week another wall went up. At the end of four weeks, we had four walls. And we had an interesting floor. Since we built on unleveled ground, the floor wasn't level either. All the water and spills ran to the low corner. That would make cleaning up after five small children easy enough. Now all we needed was a roof.

As the building project moved into cold weather, we made a special detached hallway out to our camper. We still slept, cooked, and ate in the camper. Mornings were interesting and invigorating! That cold floor jolted us awake faster than water in the face.

Don came home only to find out he was being sent to Nome in two days to start another job. We all worked double-time to get a plywood roof on the shack and to cover it with tarpaper. A brand new propane stove became our first source of heat, and we had an outhouse 150 feet away by the time Don left for Nome.

Our water came in five-gallon jugs from the university's fire station. We still didn't have electricity, and two days

33

later the temperature dropped to the cold weather that Alaska is famous for.

Don had only been in Nome for seven days and the car froze up, we ran out of water, and our two youngest, Jon and Susan, got pneumonia. Thank goodness a friend stopped by and took the kids to the doctor.

I'd had all I could take. Two days later five boxes were packed. Our five kids and I deserted Mom's tarpaper shack and boarded a plane to Nome for the duration of Don's job there.

Sourdoughs are the pioneers and prospectors, the residents who have spent most of their lives here. The name came from a yeasty starter, used to leaven bread and pancakes.

Cheechako, (pronounced chee-cha-ko) is a newcomer to Alaska, a tenderfoot or greenhorn.

The Fourth of July, Alaska Style

Don Elbert

Like the rest of the United States, July Fourth is a special holiday for Alaskans. However the manner in which we celebrate is a bit different. We began making plans in mid-June for an overnight riverboat trip for the Fourth of July weekend. My wife Elaine and our five children Dave, Diane, Kenny, Karl, Doug, and I talked about our expectations, each one sharing their own ideas to make this a memorable weekend. Manley Hot Springs, a quiet settlement of less than one hundred people, was our chosen destination,. Manley sprang up during the peak gold mining days of the early 1900s a few miles off the Tanana River and had been a busy trading center ever since.

On July 3, 1972, all the elements were in our favor as the sky shone brightly with only a few misty clouds lurking lazily about. The warm rays of the summer sun, plus twenty-four hours of continuous daylight, made it the ideal time for a family outing. The temperature was in the low seventies as we left our home in Fairbanks early that Saturday morning. I towed our twenty-foot flat-bottomed riverboat sixty miles south to Nenana.

I backed down the boat ramp into the Tanana River until the boat floated off the trailer. Dave, our oldest, proudly held tight to our boat from shore until everyone had jumped aboard. The children were clamoring for their own favorite niche to snuggle down and enjoy the ride. Smiles and excitement overflowed from our craft as we pushed off from the bank, and began our downriver adventure to Manley Hot Springs. The Tanana River creeps along at seven miles an hour, and it was like having a gentle wind at our back. I focused on avoiding the floating debris, logs, and trees that cluttered the river and keeping the boat off the sandbars that lurked inches under water. I stayed in the deeper water where the stronger currents flowed by, following the undercut riverbanks.

Elaine served as navigator, reading the map and using her keen observation skills to help keep us in the deep water. Frequently we had to switch from the left to the right bank because that is what the current of the river did. We followed Mother Nature's path of least resistance. We all enjoyed the scenery and the ever-changing river in our own way. I studied the river and was intent on keeping us out of danger. My wife enjoyed the vegetation, the sky, and the wonder of God's hand in it all. The children were all keen-eyed looking for wildlife.

Thirty miles brought us to the often-flooded and now deserted village of Old Minto. We pulled up to the high bank and stopped. David, my eldest son, grabbed the rope and tied us to a tree root. We climbed up the bank and walked through the remains of the deserted village. There were a couple dozen cabins. Most of the doors and windows had been removed. We found the village's schoolhouse, which was open for exploring, and our children had a lot of fun rummaging through several layers of wet schoolbooks that lined the floor. In one of the open log cabins we startled a porcupine eating something that looked and smelled disgusting. I'm not sure who was more startled, the porcupine or us, but it left first, with its quills erect, through a large hole in the floor.

After an interesting two hours of exploring the ransacked remains of this deserted sliver of Alaska history, we returned to the boat and continued our river journey. We passed several duck families out for their afternoon swim. The mother ducks were teaching ten little ducklings the art of paddling and the safety of staying close together.

A few minutes later Doug, our youngest, was the first to spot a moose swimming across the river with just its head and yard-wide antlers above water. Approaching within fifty yards, I stopped the engine to watch. I've seen moose by the hundreds, but I've never gotten over the enjoyment I feel while observing these big animals. A quick glance back at the smiling, intent looks on the red-cheeked faces of our five children revealed their excitement as well.

Exposing Alaska

We watched this moose until it reached the bank, where it slipped and stumbled and on the fourth try was able to climb up on solid ground. Then, like a ghost, it vanished into the spruce trees.

Around the next bend, Ken, our middle son, saw two beavers out rebuilding their dam, repairing the damage to their home caused by the spring ice breakup. Karl asked, "Dad, why do beavers always slap their tail before diving under water?"

I answered, "Since beavers don't have telephones, they slap their tail to signal to their mate it's time to get lunch ready; they'll be home soon."

Karl replied, "Oh, Dad."

It was early afternoon when we left the Tanana River and entered Manley Slough, which leads the last seven miles into Manley. The kids were anxious for a break after an invigorating morning on the river. We docked at the campgrounds just outside of town. There was a large gathering of local folks there. We had arrived just in time for the annual Fourth of July celebration. The bountiful picnic featured roast duck, several varieties of salmon, and a big pot of moose stew. I was in the process of devouring my bowl of moose stew when the air show started. Seven paratroopers in brightly colored jumpsuits bailed out of an airplane and floated down, each landing on their feet inside a twenty-foot circle in the middle of the baseball field.

The races and games that followed were delightful. Diane, our eleven-year-old, was a cheerful girl with bouncy golden curls before she entered the blueberry pie-eating contest. She didn't win, but when she looked up her face had turned a deep dark purple, the color of a juicy

∽ Exposing Alaska ∽

over-ripe blueberry. The purple shine on her cheeks stayed with her all day as her badge of participation. Even though her curls had lost their bounce, she continued to wear her delightful smile. I think she had more pie smeared on her face than went in her mouth.

The boys entered all the running events and even won several. Doug won the hundred-yard dash for the under ten-year-olds. Dave won the gunnysack-hopping race. Karl got about halfway before dropping the raw egg carried on a spoon in his mouth. Doug and Ken came in second in the three-legged races. Ken said they could have done better, but Doug's legs were several inches shorter than his. We couldn't have planned a more delightful day.

As the festivities wound down, we set up our tent under a lone cottonwood tree on the edge of the campgrounds. Ken tried his luck at fishing and moments later screamed for help when he caught an eighteen-inch pike from the back of my boat in the Manley Slough. Ken said, "Hurry, Dad, this fish is trying to pull me in."

Later, the kids jumped off the railing of the narrow bridge into the warm water of Manley Slough and swam around, having a grand time. They were toweling off when the skies lit up with the fireworks display. Thirty minutes of loud sizzles, pops, and bangs awakened the night as the various fireworks exploded, streaking and sizzling through the heavens with multicolored delights. However, it's harder to see the fireworks when the sky is light all night.

As the last of the fireworks faded away, fatigue set in. Our five children crawled into their sleeping bags that night happy, tired, and contented little campers.

In the morning as the sun greeted us with a warm embrace, Elaine cooked Ken's pike for breakfast over our

Coleman gas stove. Before noon, we broke camp and slid down the bank of the slough to our boat for the return trip home. All of us waved a grateful goodbye to those standing by, watching us depart. My forty-horse motor pushed us out of Manley Slough to the Tanana River, where we turned up river.

Once we were away from Manley, I still had the July Fourth fireworks theme in mind. I allowed Dave, just thirteen, to carefully take pot shots at the floating logs and debris in the river using my .22 pistol.

I found myself lost in the beauty as I enjoyed the calmness and tranquility of the Tanana River as it meandered through the fifty-mile wide Tanana Valley. The glistening, snow-covered mountains of the Alaska Range dominated the background. Above the mountains was a crystal clear blue sky, like a giant dome, covering all God's creations. The sun was at our back as we joyfully tooled back up river toward Nenana. I turned around to capture my family's mood as the peace and serenity of this excursion swelled in my heart. A smile crept across my face as I looked upon the fully alive and contented faces of my wife and each child. A close look into each smiling face was all the proof I needed to know that they were enjoying our outing. Elaine flashed me her biggest, most beautiful smile (the one I fell in love with) which told me this July Fourth had been very special for her too. Diane and the younger boys were happily eating potato chips.

My full attention centered on Dave when I noticed he was looking skyward and his gaze was slowly descending. I glanced up in time to see a raven falling into the river. That little rascal had shot a raven in flight, dead out of the sky.

That ended the target practice. But I did tell him, "Son, you're getting to be a good shot."

We made good time back to Nenana, loaded our boat on the trailer, and returned home. Reflecting over the last two days, I thought: "This riverboat excursion and togetherness with my family has been a hallmark experience. What an extraordinary way to spend the Fourth of July in Alaska."

Alaska is a geographical phenomena: active volcanoes, frozen glaciers, hot springs, permafrost, the deepest lakes and the highest mountain in the Western Hemisphere.

The three bodies of water that make up Alaska's 33,904 miles of coastline are the Arctic Ocean, Pacific Ocean and the Bering Sea.

The largest gold nugget ever found in Alaska was discovered on September 29, 1903, in Nome. The nugget, weighing 155 troy ounces, was seven inches long, four inches wide, and two inches thick.

Seven Moose on the Trail

Cathy Webb

Bushels of yellow leaves floated down, exposing scraggly naked tree branches. It was late August in Eagle River and my husband Ron and I were out doing our three-mile walk on the loose golden carpet. We still had a mile to go and I was leading the way when suddenly fear stopped me dead in my tracks. A huge bull moose stood menacingly right in the middle of the trail. Six other anxious bulls were close by, and these males were snorting and vigorously pawing the ground. This was mating season and these seven animals had waited all year for their chance. We had walked into the arena where these bull moose were going to fight for dominance.

Ron quickly pushed me off the trail so I wouldn't be trampled or hurt. That big moose wasn't about to move from the narrow trail. He just stood there, his large brown eyes glaring at Ron. He never even blinked. He was staring my husband down. It was already late afternoon and starting to get dark. After a long stare-down match which the moose won, we crept backward until we got around the corner and then returned the way we came.

We were in an area known to harbor two wolf packs, several bears, and loose moose on the trail. To make matters worse, I suffer from multiple sclerosis and I was getting over tired. We sang songs and talked loudly to let the animals know we were there. We tripped over protruding tree roots on the trail in the dark. It was near midnight when we completed the two-mile walk back to the car. What started out to be a nice quiet hike had turned into a frightened ordeal. I've had bears rummaging in my garbage cans on the front porch before, but I was safe in the house. It's really startling when you confront seven big moose right in front of you on the trail and know there could be a serious confrontation at any moment.

The moose is Alaska's official land mammal. Moose are the largest member of the deer family. Full-grown males stand six feet tall at the shoulders and weigh from 1,200 to 1,600 pounds when in prime condition. Moose are long-legged and heavy bodied with a drooping nose, a "bell" or dewlap under the chin, and a small tail. Bulls can carry antlers that weigh up to one hundred pounds and measure up to seventy-two inches in width.

I Held onto My Babies and Watched Our Belongings Float Downriver

Ann Dolney

Frontier life on our own piece of ground back in 1957 was rewarding us for our hard work. My husband Ed and I had homesteaded our 160 acres eight miles west of Fairbanks. We cleared the land, built the house, jetted the well, and planted the fields. We were also blessed with two beautiful baby girls we named Linda and Karen. We named our domain Happy Acres; it seemed appropriate, since we were very happy there. I loved my pioneer life and didn't ever want to leave our home.

However, in order to pay the bills, Ed's day job was working for the Alaska Road Commission. His new assignment was to widen and expand the runway in the remote village of Wiseman so the larger cargo planes could land there to bring supplies. This sparsely populated village was only accessible by air in those days.

The thoughts of Ed being away for three months overshadowed not wanting to leave our homestead. So I packed our bags and bundled up our two baby girls. Frontier Air Service flew us to the nearly abandoned mining town of Wiseman. We set up housekeeping in a cabin owned by Joe and Tissue Ulen. Wiseman is a picturesque and beautiful spot tucked high up in the Brooks Range. We spent many happy hours watching bands of sheep grazing and cavorting on the rocky cliffs of the surrounding mountains.

One of Alaska Road Commission's D-8 Caterpillar tractors was left in Wiseman for clearing snow in the winter. Ed is an experienced mechanic as well as a cracker-jack operator. He had to replace several of the worn track rollers and a leaking water pump. He adjusted the sagging tracks to get the D-8 in working condition. Ed's keen eye and steady hand on the cat's blade enabled him to widen and lengthen the runway in record time.

His next job assignment called for a move to Porcupine Creek, fifteen miles away. We loaded all of our belongings on a low four-by-eight-foot sled, which we pulled behind the D-8 tractor. Ed had built a special wooden box for our two little girls to sit in. I lined the box with a blanket and added Linda's favorite teddy bear before placing the box on top of our belongings in the sled. Karen was clutching a box of Oreo cookies as Ed lifted her and Linda into the choicest seat on the sled.

Ed made a final check of our load and determined it safe and secure. He then tickled Linda under her arm and pulled Karen's ear to make the girls laugh, and Karen handed him one of her Oreos. With everything in place and an Oreo cookie in his mouth, Ed climbed into the driver's seat. After starting the noisy thing he guided the D-8 tractor pulling the sled and us down the trail heading for the village of Porcupine. I rode beside the box that held Linda and Karen.

The warm rays from the sun overhead brightened the day, and I enjoyed the scenery. We couldn't talk because the loud diesel engine drowned out our words.

∽ *Exposing Alaska* ∽

We had to cross the middle fork of the Koyukuk River to get to Porcupine. We had no problems crossing the seventy-yard-wide, shallow Koyukuk River the year before. But this year more snow and ice melting in the mountains had caused the river to rise. To my horror, in the middle of the river, the sled began to sink. The water came up so fast I didn't have time to think, I just screamed for Ed. I felt the chill of the glacier cold water rising up my legs and I turned to check on my babies. The box with our two little girls in it had started to float! I grabbed the box with one hand and held onto the railing on the sled with the other. Ed couldn't see what was going on right away and didn't hear my screams over the blasted engine noises. I was holding onto the box with our girls in it for dear life. I watched all our belongings—hip boots, sleeping bags, fishing gear, clothes, boxes of food, everything that would float—drifting down the river. Ed finally turned around to check on us and gasped at our predicament. All I could do was motion to him with my head to keep going and get us across the river. When we came up out of the water I was cold, wet and shaking so badly I could hardly stand.

We had lost most of our gear but the supplies were replaceable. My main thought was we hadn't lost our girls the way we lost our belongings to the Koyukuk River. My husband jumped down from his high seat and we all hugged tightly together without saying a word.

Which One Jelly

Ann Dolney

Linda, Karen and I used to go behind the cemetery in Wiseman and pick currants, and I would make jelly. One day our neighbor, a Native lady named Ester who spoke broken English, stopped by with her daughter Louisa and tasted the jelly. She wanted to know how to make it taste so good. I told her, "I will show you how when we pick more currants."

She said, "Let's go pick some more now."

So we did and after we had our pans full of currants we returned home. Then I explained to Ester how to make jelly,

Ester making jelly at Wiseman.

and I showed her how to put the berries in a sack and squeeze the juice out. The next morning I was busy in the house but I could see out the window that Ester had the tripod up with a pot on the fire and the fire going. All of a sudden Louisa ran over and said, "Ann, Ann come." So I went over and Ester was standing in front of the fire with a pot full of currant juice and in her hand was a sack of pulp. Ester looked at the pot and at the sack and asked, "Which one jelly?"

When we moved to Wiseman we had to take all our supplies, canned or dried foods, flour, sugar and everything else we would need. I included a sack full of rock candy treats for the girls. One day I gave Louisa some rock candy and she seemed to enjoy it very much but never really said anything.

Several years later, when she was a grown woman, she came to Fairbanks to visit us. Louisa said, "Ann, I sure remember you from Wiseman. After you left I must have bit into every rock along the Koyokuk River trying to find some more of those rock candies."

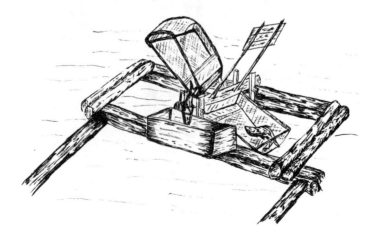

A Lonely Birch Tree

Don Elbert

When I was a boy, one white, lonely birch tree stood erect like a sentinel guarding our homestead. A scant thirty yards off the dirt trail that was Farmers Loop, out in the middle of the cabbage patch, this tree overlooked the winter ice fog down in the valley. How this birch matured to its height of thirty-five feet is a mystery to me. Everyone knows that trees don't grow when it's forty below.

This tree was special because it was the only tree on our farm to survive land clearing. Dad had rid the land of all the other trees in his ongoing efforts to increase the acreage for the money crops of cabbages and potatoes. I felt sorry for the tree, thinking it would get lonely because it didn't have any tree friends around, and sad because all its neighboring trees had been chopped down and sawed up to appease our ravenous Yukon stove.

The tree's scrawny naked branches came alive in the spring; new buds tentatively peeped out for a glimpse before bursting open into green oval leaves. Swallows would land on the thin branches and hide amongst the leaves. They scoured the freshly disked earth below, searching for tasty grub worms.

Those same leaves also gave shade to my brothers and me when we took a break from hoeing the cabbages on a hot summer day. But the green leaves didn't live long. In a few short months they turned brown and died and gradually fell to the earth.

The savage winds would come, pummeling the tree, blowing the last of the protective leaves away as the killing frost tiptoed in like a thief in the night. Stripped of its leaves, this lone tree stood nearly naked, wearing only its thin white bark jacket for protection against the cruel and deadly cold.

Several days later the heavens opened and snow came down by the buckets full, enshrouding this birch tree with a snug warm white blanket. That night, I heard the tree creaking under the weight of the freshly fallen snow. In the morning the tree was bent, like an old man struggling under the weight of his years. Once I tried to count the snowflakes balancing on one of the skinny limbs but couldn't, The flakes all clumped together as one.

Occasionally a traveling moose ventured by and alternated between digging up cabbage leaves from under the

snow to dancing around this birch tree and eating all the twigs from the branches as high up as it could reach. Dad's Farmall tractor, exposed to the elements, sat dormant in the winter shadow of our tree. The thick layers of snow offered no protection to the metal parts of Dad's hibernating workhorse.

Upon returning home in May from a winter in Barrow, my daughter-in-law Kelly said, "I love looking at that birch tree, seeing the green leaves coming alive on every branch. I miss that. There aren't any trees in Barrow."

Over the years I watched this tree overcome all obstacles. It stood tall against temperature changes from ninety above and the twenty-four hours of sunlight in the short Alaska summers to sixty below in the long cold dark winter months. But nothing lasts forever and trees are no exception.

The years caught up to my Dad: he's gone now—the farm too. The acreage has been divided and is now Husky Garden subdivision. That lone birch tree isn't there anymore either: last year it fell from the weight of too many snowflakes. On a forty-below-zero night, in my efforts to keep the cold at bay, I cut that dead birch tree into two-foot lengths and respectfully began cremating it in our insatiable Yukon stove.

Alaska—The Best Deal by Far
The bulk of Alaska is owned and managed by the federal
government, which purchased Alaska from Russia in
1867. (This purchase for $7,200,000—or less than two
cents an acre—was by far the best bargain of any land
purchase deal in U.S. history.)

Dad and I

Don Elbert

My 260-pound dad, Paul Angelo Elbert, was a giant among men in size and ability. His erect stature was a visible expression of his pride and confidence. Over the years my father has continued to be my inspiration. Dad carved a hundred-acre farm out of his scrub tree homestead on Farmers Loop. He named his farm Husky Gardens, because he grew husky vegetables: "bigger and better than the competition," he bragged. During each of his peak growing years, 1948– 1958 he harvested eighty acres of prize-winning potatoes. And for years he held the unofficial title of Alaska Potato King and Champion Bear Hugger. (The Bear Hugging award came from close encounters with tame women, not wild grizzly bears.)

However, thirty years of potato farming in Alaska had taken a toll. Dad's earthly body was physically spent, but his mind continued producing golden crops of poetry. His large blue eyes maintained their cheerful sparkle on his round beard-stubbled distinguished face. His thinning gray hair, often straying, seemed to crown him.

Now, Dad spent the major portion of each cold winter day writing and rewriting poetry in his crowded one-room cabin. He would sit motionless for hours in his homemade diamond-willow rocking chair, simply rocking and thinking. A writing tablet lay on his lap, and poised over it, ready for action, a sharpened pencil. In front of him was a small table that held more pencils, erasers, a pencil sharpener and his stained, half-full cup of luke-warm black coffee. Beside him glowed a wood-fed Yukon stove.

Dad's mind often wandered—seemingly lost in a huge warehouse of words. He would scan his expanded memory banks, searching for just the right word or phrase. He completed his labor of love in 1963, and published his first and only book of verses, entitled *I Love the Land.*

Gail Haynes critiqued Dad's work and had this to say about him: "To those of us who know Paul Elbert the man, Paul Elbert the poet seems almost unbelievable. He is slightly larger than life itself. A bluff giant of a man, whose hearty manner is only a screen behind which the poet's eye observes and the poet's heart interprets; the hands that can crush a rock can delicately hold the butterfly thought of a poem."

"Paul's book of verse proclaims his fierce love and kinship for Alaska. It was most fitting that he chose the title I Love the Land. Many of his poems are direct outgrowths of his life and loves, his pursuit of happiness and his understanding of God."

Dad was called home to the heavenly harvest of souls in 1972. I miss him greatly, but I have not forgotten him. His poetry is his legacy and is a gentle reminder of Dad. As I reread Dad's poetry today it brings back memories so clear that I can again picture Dad opening the cabin door each morning to greet the world.

A gentle smile rippled across Dad's face as he took in a deep breath of the cool, crisp morning air. His chest swelled as he inhaled the fragrance and the freshness of a new day.

∽ *Exposing Alaska* ∽

The sun crept upward, peeking over the mountaintop, adding sheen to the snowy white horizon. Dad raised his eyes to greet its brightness with reverence. The sun's rays met his eyes and touched that part of his soul which this great land had awakened in our forefathers. I sensed it was the very same awakening that created the inspiration, vision, and vitality to forge onward and carve out what we now call the history of Alaska. This daily ritual transformed Dad within moments and gave him the strength and the courage to overcome any obstacles this raw land had to offer.

As Dad's middle son and many years later, I still live on a piece of that old farm. I'm discovering more about Dad from his poetry. I've learned that a man reveals who he is and what he is all about through the subject matter of his poems. I've become aware of many things my dad never shared with us kids; his hurts and his dreams are revealed through the words he chose in his poems.

Dad wrote about his passionate loves and the sorrows of his three failed marriages. He wrote about his tireless pursuit of happiness and about the sacredness of sex. He wrote about his fascination with Alaska, about his struggles with growing old. He wrote about God.

When I was younger, I didn't have an interest in farming or an appreciation for poetry or words. I was busy with my own life, building a house for my wife and growing family of five children and working as a heavy equipment mechanic.

Now I'm on the downhill slope of life myself. Our kids are grown, I'm retired, and now I've developed an interest in words. Words have become exciting to me, as they were for Dad. I've learned the power of words and have demonstrated that warm words spoken in tenderness can melt

54

a cold heart. Words give light to my ignorance and love, and caring words will brighten anyone's day. I have learned to control my tongue because my words have hurt others. I enjoy reading well-written words; it's like having a delightful dessert after a gourmet banquet.

A good portion of my exercise nowadays is mental gymnastics, shuffling ideas and phrases around in my head until they ripen. I suspect I inherited that trait from my father. The computer keyboard has become the tool I use to select and type the chosen words. As the words are carefully rearranged, they take on new life, and I hope they'll enlighten my family.

A moment after I tap the print icon, my legacy is printed out: a document my wife, Elaine, and my children and grandchildren will read and through it come to understand just what makes me tick. I'll close my article by telling my wife and children the important thing I didn't say often enough in life: "I love each one of you very much."

"Be of good cheer and may God guide you on your path. I'll be leaving soon to spend eternity with Dad, my brother Dennis, and my son Karl. Don't hasten to follow but be assured, we'll keep the light on for ya."

∽ *Exposing Alaska* ∽

Selections From Dad's Book, *I Love the Land*

I Love the Land
I sit upon a stone
On the hillside, quite alone
And I view the land I own
In awesome reverie.
I've learned to love this land,
This mellow, fertile land.
I think it is His hand,
His hand outstretched to me.
And I caress His hand
When I cultivate the land,
And then I understand
How He provides for me.
There's grain a hundredfold
And peaches bright as gold;
And livestock, young and old,
In His hand for me.
I cannot see His face
His voice is lost in space,
But like an earthen vase
His hand encircles me.
So I caress His hand
As I cultivate the land,
And I think it's grand
How near He is to me.

Age

Age should be honored,
If it is honorable.
Laughed at, if it is funny.
Pitied, if it is wasted.
Despised, if it is useless.
Age ripens the peach,
Colors the apple,
It should mellow the man.

From Which Tree?

In these woods
I've walked
And worshiped,
There is the 'tree of life,'
And yonder,
The tree of knowledge
Of good and evil.
I have splinters
Festering
In both hands.
I wonder
From which tree
My cross
Will be hewn!

Poetry

Poetry
Is truth,
And beautiful thoughts
All dressed up
In Sunday-go-to-meeting
Words!

Poetry

After love letters comes
poetry
As the most exotic
And exciting form
Of communication.
Secrets shared.

My Faithful Dog Team

On a Cold and Snowy Night

Richard M. Barnes story
Narrated by Don Elbert

It was a race against time back in January of 1925. Twenty mushers traveled 674 lonely miles from Nenana to Nome in less than 128 hours. They successfully delivered 300,000 units of life-saving serum to the diphtheria-epidemic-threatened, isolated village of Nome. The mushers braved the wilderness and horrible trail conditions during a raging blizzard. They crossed the treacherous frozen water of Norton Bay* in sixty-below-zero temperatures. All efforts to keep the serum from freezing failed, but when thawed it retained its effectiveness, and in a few days that serum brought the diphtheria epidemic under control. Leonard Seppala and his lead dog, Togo, became famous for their part in that memorable relay. That event inspired the Iditarod Trail Sled Dog Race of today.

Richard Barnes grew up a country boy in northern Minnesota, hunting grouse, camping, and fishing for brook trout. He completed a tour of duty in the Air Force as an air traffic controller and then returned to school at Bemidji State College on the G. I. Bill. He dropped out of college to trap beaver in northeastern Minnesota.

---

* Norton Bay is located in the northeast corner of Norton Sound.

Exposing Alaska

In the winter of 1973–74, Bill Taylor from Fairbanks, Alaska, came rolling into Bemidji, Minnesota, to race his dogs. Richard met Bill at a friend's house and told Bill how much he wanted to come to Alaska to work. Bill offered him a job and also wanted him to haul some training equipment back up, but Richard knew his old truck wouldn't survive the Alaska Highway. Bill lined him up as a dog handler for a millionaire dog racer named Merv Hillpipper. A month later, Richard quit his job in the iron mines, climbed into Merv's two-ton truck, and headed north. They drove the many miles of highway into Fairbanks.

Richard moved to Manley Hot Springs with the Taylor family in 1974 to fish commercially. After the fishing season closed he began driving dogs and trapping. He couldn't get enough of the outdoor excitement to feed his adventurous spirit. This raw and eager twenty-seven-year-old began his first winter in Alaska living out his childhood dreams.

Bill Taylor gave Richard nine yearling sled dogs to train over the winter. The only other team in Manley was a team of big beautiful malamutes belonging to his friend, Mark Woster. Mark and he volunteered to take part in the reenactment on the fiftieth anniversary of the original Nenana to Nome life-saving relay. They would carry the serum to the village of Tanana. It wasn't really serum. It was a commemorative cachet,* but they took the commitment to deliver seriously.

Two days before the cachet arrived in Nenana, Mark loaded up his sled with a wall tent, food, and a wood-burning stove plus rations and supplies for their dog teams. He

---

* ca-chet (ka-shā′) a stamp or official seal on a document.

59

would go as far as Fish Lake, set up a camp and wait for Richard to bring the serum so he could relay it to Tanana.

The commemorative cachet arrived at the starting point in Nenana by rail from Seward. There, ready and waiting was Joe Reddington, Senior, with his primed sled dogs. Joe took possession of the cachet, and the relay began with Joe running the first leg of this reenactment of the original race against time. Fifty-two miles later Joe gave the cachet to Red Fox Olson who waited at Burk's cabin near the mouth of the Tolovana River. Red in turn brought the cachet to Richard after waiting for the weather to clear up. Deep snow and lack of packed trails kept Red Fox from getting into Manley Hot Springs until late on January 28, 1975. He advised Richard, "Wait until morning and get someone with a snowmachine to break a trail."

Richard told him, "Earlier in the day Frank Gurtler and Bob Lee went out on snowmachines to check on Mark Woster at Fish Lake but had to turn back on the first hill on Tofte Road because of deep and drifting snow."

Even though this race was only the reenactment of the serum run, Richard was quite perturbed at Red Fox for having stayed at the Burk's cabin so long due to weather and poor trail conditions. Richard's sled was packed and ready. He had waited two days before he could do his part. With the help of neighborhood kids, he harnessed and hooked up his team of young huskies to the sled within minutes and then he pulled the snow hook. His fresh team ghosted away into the heavily falling snow. The temperature was a winter-warm twenty above zero as he began the twenty-eight-mile trek to Fish Lake.

Wind-driven snow crystals obscured his vision, and combined with the blackness of the night, it became

Exposing Alaska

difficult to even see his dogs. He couldn't see the lead dogs at all. They managed to get through town and out on the Tofte Road only because that's the route they went on their daily training runs. They knew the way and Richard hung on for the wild ride! He figured this adventure to be a real kick and hoped to pull into Mark's camp long before the mid-morning sun appeared. Little did he know of the difficult trail conditions and the hardship that lay ahead.

Story has it that a few years earlier, a thoughtless berry picker had left a smoldering cigarette, which later ignited a forest fire that burned most of the area between Manley Hot Springs and Fish Lake. A few blackened stubs and short shoots of new growth were all that remained. No trees were left to block the wind. The drifting snow covered all trace of the trail to Tofte and beyond.

About five miles out of Manley Hot Springs, Richard came to the area where the snowmachine trailbreakers had given up. The dogs plowed through several feet of fresh, loose snow and floundered along for a few hundred yards. Richard knew from reading Jack London, Robert Service, and others that when the snow became too deep for the dogs, the musher went ahead of his team on snowshoes. Within minutes he had strapped on his big trail-breaker snowshoes and shuffled to the head of the team patting and petting the dogs as he went by. In turn the dogs wagged their tails at him. He started uphill into the storm, guided by small shrubs growing in the ditch. Twenty-five feet at the most he crashed forward, each new step more difficult than the last. What the heck! A quick look revealed seven of the nine dogs standing on the snowshoes with him. The eight and ninth dog would have been standing there also if they could have found space. He lashed the snowshoes

61

back on the sled and unclipped the tug lines from most of the dogs. As he struggled uphill through waist-deep snow, most of the huskies crowded about his feet while the two black dogs pulled the lightly loaded sled a few feet behind.

The farther he got up the hill, the harder the wind blew. In an effort to find the trail he pulled out his handy pocket flashlight. It gave only a feeble glow. The cold had sapped the energy from the batteries. He couldn't see much, but now and then he could feel the road slope into the ditch. At the top of the hill the wind nearly blew him over. He crested the hill and fought his way down the far side. Somewhat sheltered from the wind in the valley, he hooked up his team again since they could make out a minor trail indentation in the snow. Slowly but surely they were putting the miles behind them.

The next hill was higher and the trail conditions grew even worse; without the dogs he would have panicked. These wonderful animals kept him in control of the very real fear of losing his way. After what seemed to be hours of struggling, the trail started to drop down into the valley.

This cruel weather could be life threatening, and the possibility of getting lost in the vast white wilderness unnerved him. Snowflakes, carried in on the shrill winter wind, obscured the trail and stung his eyes. Richard was standing on the narrow sled runners behind his nine trotting dogs when he made out the stunted forms of a spindly grove of snow-plastered black spruce trees. Those white, still forms brought a ray of hope. They were the first things other than the dogs and sled that he had seen in the last few hours. Seeking shelter, he stopped the sled in the grove for protection from the blowing snow. He unharnessed the team and staked them in a semicircle and gave each hungry dog half of a dry salmon. With the chores completed, he removed his parka and peeled down to his long johns and burrowed his tired body into the down sleeping bag, pulling the canvas sled tarp over him.

All alone in a frozen land, Richard lay cocooned snugly in his sleeping bag while the storm raged around him. He began to wonder Why did I ever agree to take part in this relay? He dozed fitfully off and on for a few hours, thinking, If this storm is still rampaging come daylight I'm going to stay right here.

Hours later, sensing that the blizzard had abated, he pulled back the tarp and peered into the darkness. The storm had broken and the wind slowed to a breeze as a few stars poked through the loitering clouds. Remembering his mission, he jumped up and quickly repacked his gear and hooked up the team. By the diffused moonlight, he attempted to find the way to Fish Lake in the valley before him. The trail remained buried under deep snow but he could make out where it led, and by noon he passed the abandoned mining town of Tofte.

As the storm receded, the temperature steadily dropped. Around 1:00 PM, Frank Gurtler caught up to him on his snowmachine. At last, someone to break trail, he thought. Frank poured him a mug of steaming hot, sweet black tea from his thermos and handed him a bag of his mother's wonderful smoked king salmon strips. As they sat on his snowmachine eating and talking, Richard got cold. Frank told him the temperature had dropped to minus thirty degrees. Richard could hardly believe it. Last night when he had left Manley Hot Springs in the storm the thermometer had read twenty degrees above zero. However, when the skies clear in the interior of Alaska the mercury drops

63

rapidly. He dug through his sled bag to find warmer clothes to put on under his blue denim dog driver's parka.

Just as Frank was about to start his snowmachine, two Tanana men on snowmachines came up the trail from Fish Lake. They told him he had another ten miles to go to get to Mark's camp and that Mark was worried about him. Frank decided that the trail did not need more packing so he followed the guys from Tanana back to Manley Hot Springs.

Alone on the trail again, Richard rode the runners, pedaling with one foot to push the sled and to generate some warmth through exercising. The trail conditions remained soft and slow. The low angle of Arctic sunlight turned the tips of the birch trees a beautiful shade of rosy red. He began to smell faint traces of wood smoke in the cold frosted air. They pulled into Mark's camp at four o'clock, just as it started to get dark. The thermometer taped to the handlebar of Mark's sled read minus forty degrees. After tethering and feeding his team, he cut armloads of spruce boughs for each dog's bed. They curled up, tucked their noses under their tails and fell asleep in short order. They had put in a long, long day.

Mark had dinner cooking on the wood stove in his ten-by-twelve-foot wall tent. Richard enjoyed the pure luxury of lounging in the radiant heat of that little Yukon stove, his cold hands cupped around a steaming mug of hot coffee. He found out in the days to come that as long as the stove received wood every hour the tent stayed very warm. As soon as the fire died down, the cold would rush in through the thin canvas.

Mark had fed his dogs when Richard fed his. Mark waited two hours for them to digest their food before taking the "serum" on to the village of Tanana. While they waited the moon came up, bathing the night with its soft gleaming glow. When the two hours had passed, Richard helped Mark put the heavy leather work harnesses on his big malamutes and hitch them to the big freight sled. Mark stowed the cachet in his sled bag and off across Fish Lake he went. The long shadows of Mark and his team were easier to see on the moonlit lake ice than their actual figures. In a few minutes they disappeared and Richard stood alone again; the farthest that he had ever been from human habitation in his life.

Mark remained away for three nights. While Richard waited for him to return he cut dry spruce for the Yukon stove's insatiable appetite. He tried to coax a longer burn time by adding green birch and cottonwood but the fire nearly died out; it clearly preferred dry wood.

On the second day his dogs started to bark. Mark is returning, he hoped. But no, four dog mushers showed up

from the Denali Park area on a winter trip to visit school-teacher friends in Tanana. They told Richard the snow machines and his dog team had packed a firm and fast trail to Manley. With nightfall approaching, the four decided to spend the night and they pitched their tent nearby.

Mark returned to camp near dark, four days after leaving. He had made an enjoyable trip to Tanana. He arrived with the cachet in the early morning hours, before any one stirred in Tanana. He finally found his friend Buster Kennedy, who took him in and thawed him out. After a short search they found the musher that had agreed to carry the "serum" on down the mighty Yukon River. Mark visited old friends, rested, and healed the skin on his cheeks and nose that had frozen during the night of his leg of the serum run. The village elders remarked to him that his dogs (AKC Malamutes) were just like the big working dogs that they had used in the olden days for hauling house logs and firewood and moving their families to seasonal camps. They even used sled dogs to pull boats upriver. The dogs would travel along the shore, pulling with a long line tied to a bridle on the boat. The driver would steer the boat with an oar or sweep. When one shore became impassable, they would pull into shore, load the dogs on board, and ferry them across the river.

Mark and Richard spent the night in the tent camp and early the next morning they mushed for home. Richard's nine little huskies flew along the trail. With four days of rest his team ate up the miles. An hour later he stopped in the shelter of an abandoned cabin in Tofte to make tea and wait an hour for Mark to catch up. Then Mark said, "Don't wait for me because regardless of the trail conditions my big dogs just trot along at the same speed."

The next portion of the trail reached up into the high hills that he had struggled over four days earlier. With the trail packed hard now, he just pedaled going up hills and hung on for the wild ride down into the valleys. He stopped a half-mile outside of Manley Hot Springs to brew another pot of tea and to wait for Mark. Then, together they trotted through town with their heads held high like triumphant heroes. Only one young lady ventured out to witness their return.

Romance Blossoms in the Cold

Sometimes Snowbirds Fly North in Winter

Cindy Caserta

"You've spent five summers up here; when are you going to try a winter?" was the question Mike asked during our first conversation. I was in Alaska for the summer again, doing some research on tourism. I was interviewing Mike for a part of my research; he wasn't supposed to be asking me questions, especially uncomfortable, direct ones like that.

"You're like the other snowbirds that fly south for the winter," he continued. "If you moved here, you could help me build the cabin I'm working on." It was tossed out as a challenge, a bluff. I explained I had a teaching job back home. He rolled his eyes as if he had heard that excuse from snowbirds like me before.

Later that day, remembering that delightful twinkle in his eye, I considered his challenge seriously. For years, I had a dream of a log cabin in Alaska and living a simplified lifestyle. I had helped reconstruct a cabin in Ohio the spring before, so I had some knowledge of building with logs. My father had died recently and had left me with this lesson: Don't put your dreams off, live your dreams now.

It didn't take me long to decide. Within six weeks, in early September, I was back in Fairbanks with my belongings. I was primed and ready to begin building.

Residents still talk about the autumn of 1992 as the "winter that hit during the fall." Eight days after I arrived in town and pitched a tent on the property, the snow hit. At the building site, twenty-four inches of snow fell in two days. I was more or less in shock and remember asking Mike, "Is this what winter is like in Alaska?" Everyone I talked with assured me it would melt. After all, it was only the second week of September and the leaves were still on the trees. Surely, the snow was just a fluke and we would get back to a normal autumn in a few days. I never saw the temperature rise above freezing after that.

Despite more snow falling nearly every day, we continued to build the walls of the cabin. I found out building in this environment was completely different from Ohio. Because the property was twenty-five miles north of town, there was no running water or electricity. I quickly learned how to peel logs by hand with a drawknife. When it was time to add a layer of logs to the walls, I would try to hold them in place while Mike would nail them to the log below using a sledgehammer and ten-inch spikes.

Not including the snow shovel, we only used five tools to build the cabin: a chainsaw for cutting logs, a drawknife for peeling, a sledgehammer for driving in spikes, a ball-peen hammer to countersink the spikes, and a regular hammer for the flooring, windows, and roof. We created windows as we built, keeping the wonderful northern view and southern exposure to sunlight in mind.

The last weekend of October brought temperatures of colder than minus thirty degrees in the outlying areas. The roof was almost finished and the cabin was nearly done. I struggled working outside in those temperatures and our work slowed as we became closer to our goal. For warmth,

Exposing Alaska

I wore a snowsuit that was big enough to not hinder my movements and bunny boots, which are large, white, Arctic rubber boots. "Bunnies," as they're called, make people's feet look as big as rabbit feet. Climbing around on the roof and logs was a real challenge with a too-large snowsuit and huge bunnies. With one or two pairs of gloves on, it became downright dangerous. Thankfully, there was enough snow on the ground below and enough padding in the snowsuit that I wasn't ever injured from falling, which happened once.

By the middle of November, we had completed the roof, installed an oil drip-stove, and began chinking any air leaks between the logs. Because of the thickness of the logs, it took several days to thaw and warm them. At night, we'd hear the logs crackle and split as they thawed and began to dry out. We made our windows out of single sheets of thick, clear plastic, because the logs move while drying. Glass windows can and do crack during the first year.

Our first Christmas we joyfully spent together in the warm twelve-by-sixteen-foot cabin. Mike's mother came up to visit and stayed with us. I was a little unsure how well three of us would get along in the one-room cabin for ten days, but we all enjoyed the closeness rather than feeling trapped by it.

Because I spent most of the fall outdoors, I always thought of the cabin as an extension to the outside. With no plumbing, we had an outhouse, so we'd go outside several times a day. Often, we'd snowshoe the property line and check out animal tracks. At night, we'd walk the snow-covered dirt road and watch the northern lights. Because we were outside so much, the cabin was a place to eat, rest,

70

and get warm; and because of the large plastic windows on all sides, it almost felt as if we were outside.

We both tend to celebrate winter, so a couple of years later, we decided to get married in our back yard in December. The rector of St. Matthew's Episcopal Church, where we attend, had never performed an outdoor winter wedding and agreed enthusiastically to marry us. Another Episcopal rector overheard us discussing it at church one day and he insisted he come along too.

Choosing a wedding outfit was a tough decision for me. I didn't want to wear blue jeans, yet dressing up in cold weather was impractical. I decided on a turtleneck under a fancy white sweater, with a long, denim skirt. I wore my bunnies for warmth and for traction on the downhill slope of our aisle. I wouldn't have won an award in any bridal magazines, but up here, a "smart-dressed" person is one who is all bundled up and prepared for the cold. Our wedding morning dawned cold and bright, though one must remember that dawn in December is near 11:00 AM. We had a bonfire for guests to keep warm; however, at one point before the wedding, I counted twenty-four people in our tiny cabin. I was glad I had dressed before anyone arrived.

We were lucky that the temperature warmed up to fourteen degrees below zero before the ceremony began; December days are often much colder. Our back yard was a true winter wonderland with the sun sparkling on the ice crystals. The spruce tree limbs were bent low with a heavy load of snow. Mike and his friends shoveled an aisle to lead us to our outdoor altar.

With everyone gathered around in a semi-circle and our dogs serving as ring bearers, we were married in a forty-three-minute ceremony. My little fingers on both hands were nipped with frostbite and the "chilled" champagne had frozen. Thankfully, a thoughtful neighbor brought us a bottle of champagne and had stored it inside the cabin during the ceremony, so we had our wedding toast after all.

Since then, we built another log room onto the original cabin, which doubled its size, then we added an upstairs over both rooms. There is still no electricity or running water; however, we do have a certain pride in living a simple life. Our motto during cold winter weather when other people's pipes freeze and break is "No plumbing? No problems." We have several propane lights, a wood stove in addition to the oil-drip stove, a battery-powered radio, and even a twelve-volt TV. We recently purchased a generator to watch football and movies on a VCR.

Looking back, those first four years of living in a one-room cabin that we built together and being married in the shadow of our cabin were the best years ever. I found strength in the land and in living a life without the luxuries I had growing up. People often ask me if I ever have any regrets of moving to Alaska. That question is as easily answered as the first question Mike ever asked.

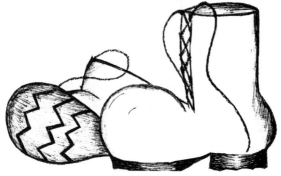

Bunny boots

Growing Up in Alaska

The Big Red Tractor

Ken Elbert

I remember feeling so grown-up as a young boy sitting on my grandfather's lap and helping him drive his big red Farmall tractor. He was teaching me how to steer straight while cultivating the endless rows of potatoes. Grandpa kept me entertained by telling humorous stories while we traveled over rows of white potato blossoms. When grandpa ran low on stories, I'd tell him about my dream of becoming a lawyer when I grew up big like him.

I was only ten years old when my giant of a grandfather died suddenly. "His heart just gave out," the doctor said. I cried, not understanding about death. A few days after the funeral and after our tears dried, my mother, brother Doug, and I went to clean his house. Of course, Doug and I weren't interested in cleaning the house, so we went outside looking for something more fun than sweeping the floor. Grandpa's red tractor, sitting in the bright sun beside the house, drew us like a magnet. I climbed up on the high seat and began cranking the steering wheel back and forth, moving levers up and down while making purring noises as if the engine were alive.

That was fun for a while, but I soon grew restless. After climbing back down, which was a long way because the big rear tires seemed nearly twice as tall as I was, I looked the tractor over. There was a wood block in front of one of the big wheels. Being smart, I decided that the block was holding the tractor in place so it wouldn't roll down the hill to the road. If the block was removed, the tractor would roll ahead. This gave me an entertaining idea. I knew the tractor would stop within a short distance if somebody turned the front wheel of the tractor uphill. I just had to try out my new theory. Quickly realizing that it would take two people to make this experiment work, one to pull the block out and one to steer the tractor, I glanced at my six-year-old brother Doug. He was the only available candidate. I didn't think he had the strength to pull the block so I told Doug he could do the steering. He giggled his acceptance and his face lit up just anticipating how much fun guiding the big tractor would be as I helped him up on the driver's seat.

I gave him an easy to understand task, "You just sit on the seat and turn the steering wheel to the right when the tractor starts rolling." I showed him which way was right, but he still looked a little confused. Then I said, "Okay, get ready," and I yanked the block. Sure enough the tractor slowly started rolling and I yelled to Doug, "Turn right, turn right!" Doug turned the steering wheel to the right, and the tractor turned into the hill and slowly came to a stop. I put the block back under the wheel and waited for my heart to stop pounding. Then I decided it would be more fun if the tractor went a further distance and got going a little faster before it stopped. So, after explaining this to Doug and having him turn the wheel down hill, I again yanked the block and gave the tractor a little shove. It began rolling. I waited a little while longer before telling Doug to turn the steering wheel and this time it went faster and farther than before. I felt pretty good about that and how my experiment was turning out to be so much fun. Twice more we tried, each time letting the tractor go farther and faster before I yelled, "turn the wheel," and he rolled to a stop. "Good job," I told Doug and his face beamed his pleasure.

The fourth time, wanting even more of this thrilling experiment, I watched as the tractor picked up speed and was going faster than before. At the last minute I yelled to Doug, "Okay, turn the wheel." I looked up in disbelief to see Doug bailing off the tractor backward. He had a look of frozen terror on his tear streaked face. I screamed at him, "Why did you jump off?" His lips quivered and the tears flowed as Doug cried, "It was going too fast. I was too scared."

We breathlessly watched the driverless tractor gaining speed as it rolled on down the hill toward the road. It began bouncing up and down and rocked back and forth as it rolled across the hills of potatoes. The tractor tilted to one side then the other so much that I thought it would surely tip over, but it didn't. It rolled faster as it headed straight down the hill towards the heavily traveled Farmer's Loop Road. Then the raw fear of what would happen if the tractor slammed into a car overtook me and I felt sick. I wanted to run away and hide. The worst feeling I'd ever felt engulfed me, and I wished this nightmare would just go away.

Luckily for us, at the bottom of the hill, there was a ditch before the six-foot incline leading up to Farmers Loop. The big tractor slowed down as it climbed that incline. It lumbered to a stop just before reaching the road, as several cars whizzed by. I breathed an enormous sigh of relief and both Doug and I ran into the house to help Mom with the cleaning. We didn't have to say anything to each other. There

Exposing Alaska

was an unspoken rule between us that neither would tell anybody what had happened. We gladly helped Mom the rest of the day, but I continued to have this sick feeling in my stomach.

That night when Dad came home from work, he noticed the tractor in the ditch at the bottom of the hill. I overheard him ask Mom about it, but she didn't know anything. The next day I was with him when he asked a neighbor if he knew who moved the tractor to the bottom of the hill but the neighbor didn't know either. It was a mystery for three days. My dad never suspected or asked me if I knew what happened, but I knew he was bothered about it. On the third day, Dad and a friend were outside and the subject of the tractor came up. Dad told his friend he couldn't figure out how the tractor got in the ditch. Finally, I decided that I couldn't live with that sick feeling in my stomach any longer.

I fessed up, "Dad, I know what happened," and I related the whole story. I concluded with, "It was all Doug's fault because he bailed off instead of turning the tractor uphill." Dad studied me with a puzzled look for a few seconds in disbelief. After digesting my story he finally said, "I'm proud of you for telling the truth son, but you and Doug could have been killed or caused an accident killing others. Don't ever do anything like that again." I did receive a soft swat for my misdeed but it didn't hurt.

The record low temperature was minus eighty degrees Fahrenheit at Prospect Creek Camp in the Interior on January 23, 1971, although colder temperatures have occurred without official verification.

My Go-Cart

Ken Elbert

One month after the tractor incident and after realizing the big tractor was not a play toy, I decided I really wanted a go-cart badly. I convinced my nine-year-old brother, Karl, how much fun it would be to race around in a go-cart so he would help me in pestering Dad about it. I knew we didn't have the money to buy one of those fancy store-bought, red go-carts made of steel tubing and powered with a real engine to make it go fast. We were fortunate however, because Dad had a strong desire to make us boys happy. So, from the junk pile (Dad called it his goodie pile) out in back of the house, he took some plywood and a few two-by-fours. Next, he removed the two axles and four rubber wheels off our wagon that a wandering moose had trampled the winter before. Then together, father and sons started building our go-cart. The steering wheel was made from plywood cut in the shape of a dinner plate and then attached to a broom-handle with rope wound around it leading to the front axle for steering. The brake was a one-by-three-inch board that pressed against the right rear tire when you pushed really hard on it with all your might.

Two weeks and many splinters later, we completed the last of the detailed sanding and finished up on the steering wheel. Then, Karl pushed while I, about as happy as a boy can get, steered our newly assembled cart out of the garage. It wouldn't have won a prize for looks and it wasn't painted red, but it was sturdy and we were happy to have it. There was no engine, but that was okay because we lived on a hillside. Karl and I took turns pulling and pushing it up our hill and riding it down. My brother and I spent many afternoons riding the go-cart on the road above our house. It had a gentle incline and we got going fast enough but not too fast. On the ride downhill, I felt alive and free while experiencing the thrill of floating through the air. Even though we sweated and strained while pushing and pulling the go-cart back up the hill, we did it over and over again.

Now, there was a much steeper hill down below our house. It was just a narrow walking path to the school bus stop, two hundred yards down below on Farmers Loop Road. The path was so steep that when it rained or snowed we would have a hard time walking up after getting off the school bus. In some places, we would have to resort to climbing on all fours. Two, six-inch diameter birch trees, one on each side of the path, loomed large at the lower end of this incline. For the seven months of winter snow, we had fun sledding down the path. The trick was to steer the sled between the trees. Most of the time we did. Few sledders were brave enough to sled from the top of the hill, since those two trees appeared too close together when viewed from the top.

My brother and I had taken our go-cart to the top of that path many times and would dare each other to ride the go-cart down the hill; however, we were too afraid (or too smart) and we never did it. Then one sunny day, Dad's grinning, redneck friend Fred came over. He brought his shy nephew, Curt, with him. Karl looked at me with that gleam in his eyes that said, "Let's see if we can get Curt to steer the go-cart down the big hill."

We led Curt outside and I began my overly friendly enticing by telling him, "Curt, it's just like driving a car. It's so much fun to ride as you coast the go-cart down hill. You'll feel like you are floating on air and all you have to do is just steer."

Karl fetched the go-cart from the garage and Curt's solemn face did not give a hint of his emotions. But I did see a small grin break the surface when I told him he could have the first ride. The three of us headed straight for the top of that steep path with an expression of anticipated pleasure growing on our guinea pig's face. Karl seated Curt on the soft pillow seat of the go-cart at the top of the hill as our excitement mounted. Then, I imparted two very simple, but important, easy-to-follow instructions before pushing him over: "Steer between those trees down there, and push the brake lever really hard to slow down." Then with a gentle pat on the back for being so brave, we pushed him over the brink of the hill. A quick glance at his face revealed his fear, but it was too late to stop.

Even now, years later, I sometimes wake up in the middle of the night recalling what followed. As the go-cart quickly gained speed, he yelled back, *"Where's the brake?"* and then a much louder scream that shouted with pure fright, *"I don't want to do this!"*

Karl and I watched breathlessly from the top of the hill as the go-cart went careening down the narrow path all the while gathering more speed. It began to clatter and bounce as it went over humps and bumps. Curt was getting closer to the trees on the right side when Karl and I yelled in uni-

79

son, *"Steer left, you're going to hit the tree."* But he didn't listen . . .

There was a loud crash and we watched in pain as our go-cart disassembled before our eyes. Wheels flew off and two-by-fours and plywood again resembled a scrap pile. Then there was silence and my heart stopped for a moment. I wondered if Curt was dead. Karl and I raced down to survey the wreckage. Curt was wrapped around that birch tree, bleeding from a cut on his head. His right leg was strangely twisted. He was crying in a low steady audible moan, "My leg, my leg's broke," so at least we knew he wasn't dead.

I ran back to the house crying and yelling, "Help, Dad, help! Curt crashed the go-cart." Dad and Fred rushed out wearing long faces and worried looks to assess the injuries. They carefully carried Curt to Dad's truck and took him to the hospital.

A week later I saw the shy, stone-faced Curt again as Fred brought him home from the hospital. He was in a cast that went all the way from his hip to the bottom of his foot. Curt gave me a vicious look. We never did become friends.

Caribou and reindeer are the same species. In Alaska and Canada the domestic forms are called reindeer. Most adult male caribou and moose are "bald" (shed their antlers) by January after the rut. The female caribou shed theirs in the spring. The difference between antlers and horns is that antlers are shed each year and a new set is grown; horns are permanent.

Experiences of a Twelve-Year-Old

Dick Gilcrest

Like most twelve-year-old boys, I wasn't looking for more work, but there was something enticing about the way Mr. Paul Elbert presented this job opportunity to me. Mr. Paul farmed the thirty-acre potato field adjacent to where I lived in a typical Alaska small log cabin with my mother, two younger brothers and my step-dad.

Shortly after the last snowflake melted last spring, I watched Mr. Paul on his red tractor going around and around, disking that thirty-acre field. I remembered thinking how much fun that would be, if only I were bigger. Well this spring I'm bigger, I'm twelve years old now, and Mr. Paul was asking me if I wanted to help him farm. He asked me, "Do you think you're big enough to drive my Farmall tractor around the field and disk the soil so I can plant potatoes?" After a quick glance at Mom and receiving a nod, I answered with an emphatic "Oh, yes I am, Mr. Paul."

Up close, that tractor appeared huge; the tires were taller than I was. It was an effort for me just to climb up to the seat. Mr. Paul was patient with me and showed me what each pedal and lever operated. He taught me how to drive his tractor when my legs weren't quite long enough to reach the foot pedals. I had to scoot forward in the seat, but I did it. I loved driving that big tractor and knowing I was trusted made me feel like a grownup. Before spring planting, I would rise early, at 6:00 AM, on the weekend and start disking, often not finishing until around 9:00 PM. Mr. Paul was like a father to me and taught me a lot about life and how to do those things on a farm that one can learn only from experiencing them first hand.

As the hours and the days passed in the driver seat of the tractor, I enjoyed the feeling of being productive and doing a man's job. I liked to turn around in my seat and see the straight and fresh furrow of dirt that the disk had just turned up. I felt proud of my uniform rows and enjoyed the warmth of the sunny spring day. Most of the time I would just let my mind wander and think about how to spend the money I was earning. I decided to save it for four more years so I could buy a car and be legal to drive on the roads.

Mr. Paul taught me how to disk in a straight line by picking an object at the far side of the field and steering straight toward it. He showed me how to cut the corners of the field and then go back and forth in order to cover all the areas I missed. I also experienced getting too close to the edge of the field and getting the tractor stuck. I felt really bad about getting the tractor stuck, and my young ego was shattered. My head hung low as I walked back and fessed up to Mr. Paul. He put his arm around my shoulders and said, "That's okay, Dick, getting stuck is just a part of farming on wet ground." He assured me that it was not that difficult to get the tractor out. He showed me how to disconnect the disk first and put logs and sticks in the ruts that the back tires had made. He rocked the tractor back and forth in order to get the tractor up on the surface again. Then he had me connect a chain to the disk after I straightened the angle of the disc blades and he pulled it through the bog.

82

My Worst Farm Experiences

Dick Gilcrest

Besides growing potatoes, Mr. Paul raised pigs and chickens for market. I was not a happy boy during pig butchering season in the cool fall. We would start that day off by filling an old cast-iron bathtub with water I had hauled from the creek. Next, it would take an hour to boil the water using a propane weed burner. Then the hog was shot in the head and bled, after which we would hoist it up and then lower it into the scalding hot water. Up to now everything was okay. Then the hog was pulled out of the hot water and laid on the table in order for us to scrape the hair off. My gross job was to help scrape hog hair, but the smell and the steam was almost more than my stomach could take. I lost my breakfast several times before I learned how to keep my nose out of the steam and to concentrate on anything other than that sickening smell.

Mr. Paul also raised around two hundred chickens for his recently opened restaurant, which he named Mom's Chicken Dinner Inn. My worst ever job while working on the farm was removing the all-winter-long accumulation of chicken manure from the chicken coop. I would gather the manure up by the forks full, fill the wagon, then pull the wagon out to the back of Mr. Paul's GMC stakebed truck. From there, I would pitch the chicken manure up on the bed of his truck and go back for another load. The manure smelled pretty ripe and was alive with little wiggling maggots. The odor was enough to gag a maggot, I thought, but I could see that wasn't true, the maggots were doing just fine: I was the one up-chucking. I would then have the privilege of walking behind Mr. Paul's truck as he drove slowly through the fields. Between gagging and swallowing hard to suppress my now-empty stomach from further efforts to revolt, I unloaded the pungent manure, one pitchfork scoop at a time, and spread it as evenly as I could.

I am older now and an official sourdough. The cold weather is definitely a new experience every year, and it does not make any difference how many years you live in Alaska.

Often, I like to reminisce back to my twelfth year of life and to the experiences I had on the farm. I learned a lot that summer from Mr. Paul about life and working in my first job. One big lesson I learned: If you want to raise chickens, you had better be willing to scoop a lot of poop.

Three of the ten largest earthquakes recorded in the world occurred in Alaska. The magnitude 9.2 earthquake that occurred in south central Alaska March 27, 1964 was the second largest earthquake ever recorded. This earthquake released sixty-three times as much energy as the largest recorded earthquake in California.

Scariest Moments by the Pilots

Flying in a Whiteout

W. T. Reeves

Our crew was pushing hard to finish our section of the Trans-Alaska pipeline first. In May of 1977, I was working out of Isabel Camp, east of Summit Lake just below the continental divide. It was a beautiful location. I had my 1975 Bronco, a riverboat, and a Cessna 150 aircraft with me. I worked nights by choice and fished or flew all day.

As a heavy-duty mechanic, I only worked when something broke down. The rest of the night was spent driving up and down our seventy-five-mile part of the line sightseeing or sleeping in the truck. We were now having twenty-three hours of daylight, and new wildlife was everywhere.

I had a young, twenty-one-year-old fellow named Bill assigned to my truck to help me. No one else would accept him, saying he was too inexperienced. I knew how eager young men were to learn. All I had to do was watch and instruct Bill and he gladly did most of the labor. He also liked to drive the truck, so I was on one long vacation, and being paid two thousand dollars a week.

One morning about 2:00 AM the temperature began to drop and the sky turned gray. I was no stranger to the area or the weather. I had been caught in a freak snowstorm back in 1957. That July 28 storm lasted for twelve hours and dumped over a foot of snow in the area. It had been two days since I had flown into Fairbanks to visit my family. Normally I would not have planned a trip again so soon, but I figured we were in for a slow drizzly rainy day, and maybe the sun was shining in Fairbanks.

We had to drive by Paxson Lake on the way back to the camp, so I asked Bill to drop me off at the gravel airstrip where I kept my plane. A quick preflight and I was on my way. My Cessna 150 climbed to eighteen hundred feet in the cold morning air. In a few minutes I was approaching the point of no return in Isabel Pass. A few snowflakes slid off the windshield, and five miles ahead the dreary sky was being painted white.

I banked to the right and flew up to Isabel Glacier. I banked to the left and looked straight down the glacier where it emptied into the Delta River. The storm was heavy enough that I couldn't tell where the glacier stopped and I couldn't see the river. I had hoped to fly down the glacier and bypass the storm, but that wouldn't work. I turned to the left, passing over Isabel Camp to return to the gravel airstrip near Paxson.

There was heavy traffic on the road at this time of the morning, so I had no trouble hitching a ride back to camp. There were several parties going on in different rooms and I stuck my head in, sniffed the air for the odor of pot before going in. Pot parties were less noisy, but boring. I only drank beer and liked to tell jokes, but people on pot couldn't keep up with the punch line. They just sat staring at me with a blank face.

When I finally found a beer party, my foreman Bob was stuttering away,

"H-a-y-y—y- W-W-W- -T-T—T, H-H-H-A-VE A-A BE—BEE-EER."

∾ *Exposing Alaska* ∾

Bob was a redneck Alabama boy, like myself. He was fascinated that an old Alabama son of a sharecropper could fly a plane, let alone own one. He often looked me up at night and talked for an hour or more about flying. Today would be no different. I had a Schlitz beer in my hand and was busy answering Bob's questions, and retelling some of my past flying experiences. A half-hour later, my cold beer was warm and still over half full. Bob had just enough to drink to put him in a happy mood. My stories had excited him to where he was ready to fly.

Bob said, "L-e-e-t-s go-g-g- t-o to Fa-iriba-ks."

I said, "We can't."

Bob said, "W-h-why n-n-n-o-t not?"

I said, "It's snowing in the pass."

Bob said, "H- o-ho-w you-you k-k-no?"

I said, "I flew out an hour ago."

Bob said, "I-t- —it q-u-it by-by-by now."

I should have listened to the little warning in the back of my mind, but I brushed it aside. I said, "Bob, if you really want to go, let's do it now, because I won't fly after more than one beer."

After driving back to my plane we took off and as we crossed the continental divide, the storm was still hanging in the pass. I banked to the right to go back and Bob said, "I-I-I- C-C- CAN S-E T-H-RRR OUGH IT."

I had told Bob my rule of thumb: If I can see through a rainstorm or snow, it is visually safe to fly through. What I hadn't told him was this rule was in wide-open places, not in a narrow pass with a dogleg halfway through. There was no way to know what was beyond that curve in the pass.

Bob insisted, and I broke the second rule: Never be talked into flying against your better judgment.

87

Exposing Alaska

As we passed the point of no return, we entered the snowstorm and I couldn't see through it. I did what I should have done before starting the trip.

"Radio, Fairbanks, 22185."

"Go ahead 85."

"How is the weather in Fairbanks?"

"We are in a white out. What is your location?"

"In Isabel Pass."

"Is the pass open?"

Not wanting to discuss my predicament I replied, "barely."

By looking down I could see the highway occasionally, just two hundred feet below. I could make out dark objects sometimes, which had to be the canyon walls. After several minutes everything went white. I told myself we had made it through the curve and could now fly instruments to Delta. I didn't want to gain altitude for I was still hoping to find the road below. If I didn't find the road soon I would have changed plans because somewhere out there, in all that white, was Donnelly Dome. The Dome was a broad, bald mountain, 3,910 feet high, south of Big Delta.

Although I wasn't instrument rated, I had logged in seven hours on instrument flying legally and probably another seven hours getting into the same mess that I was in now.

I tuned in Eielson's VOR (very high frequency omnidirectional radio range station) and set the dial to 125 degrees from Eielson. I still had one last hope, a small landing strip at Black Rapids. The radial lines are like spokes from the hub of a wheel, that get wider as you get farther from the VOR hub. But at least the 125 degrees would let me know that I was close to Black Rapids. The needle swung to the center, but I couldn't see anything ahead or

∾ *Exposing Alaska* ∾

below. As the pointer drifted slowly off center I switched to Fairbanks VOR and centered the needle. I was on the 115 degrees radial from Fairbanks. Knowing the distance and how wide the radial would be, I didn't dare push my luck; Donnelly Dome was out there somewhere. I switched to Big Delta VOR and started a slow right, watching for darkness through the snow. The Big Delta VOR wasn't giving me much help. I figured Donnelly Dome was blocking my signal. After a few minutes the needle began to swing and I knew I was east of the dome. I centered the needle and began tracking the 345-degree radial to Big Delta. I switched back to Eielson VOR and set the pointer to 120 degrees. The needle swung back and forth. I knew the dome had me blocked to my left. But again, after about one minute the pointer swung to the center and I knew Donnelly Dome was now behind us. I turned back to Big Delta and was still on the 345-degree radial. By flipping back and forth between the three I knew where I was at all times. My route would put me flying right over Ft. Greely. I picked up the mike and triggered it,

"Ft. Greely, 22185, how do you read?"

"Loud and clear, how can I help you?"

I explained my predicament and asked for permission to attempt an instrument landing on base.

"That's a negative 85." Ft. Greely sympathized with me, but couldn't take the chance on my flying into a building that could kill dozens of soldiers.

I was now down to two choices. The first was to declare an emergency and fly instruments to Fairbanks. If the controller didn't fly me into a tower it would most definitely get me grounded for ninety days. The second choice was to attempt a blind landing at Big Delta. I would have to

89

draw from all of my commercial training, and even then I was flirting with death.

Bob's face was no longer red from the beer and excitement. He looked drained and white. He quietly watched as I flicked back and forth between Fairbanks, Eielson, and Big Delta VOR. I was tracking in on the 345-degree radial at Big Delta and waiting for Fairbanks radial to center at 100 degrees and Eielson at 105 degrees.

The pointer was hard to keep centered on 345 degrees and I knew I was very close to the VOR. I also knew there was a tall tower waiting for me if I was off course by just a few feet. Even when I was dead over the VOR it would do me very little good, for the runway would be several hundred yards beyond and to my left.

The pointer swung all the way to the right, then all the way to the left, then the flag flipped from "TO" to "FROM," meaning that I had just crossed over the VOR. Now I was on my own. I felt like a blind man trying to cross a four-lane highway. It would be strictly luck from here on in.

I had never been lost in my life; I always had some mysterious feeling that seemed to tell me where I was. It was as if a picture was implanted in my brain. I imagined seeing the runway. I began counting to myself, until something said "NOW." I did a 90-degree left bank and went to full flaps. I throttled back to 1500 rpm and waited. For a split second there was black that filled the windshield. A treetop was sheared off with the prop. It hit the windshield so hard that the standby compass shot between Bob and me into the rear of the plane. I pushed the throttle in full. With the 150 horses and the full flaps the nose reared high; at the same time the tail section caught the top of the sheared off

90

∽ Exposing Alaska ∽

tree, pulling the nose level. As the tree let go I pulled back to 1500 rpm, knowing that we could be no more than fifty feet off the ground. Neither wing had caught a tree, telling me it was a loner.

Just before the wheels touched I saw gray out front and chopped power, getting ready to cut the fuel in case we flipped. There was no bounce and we began to roll down the short runway. I had landed dead on the end of the runway.

Bob jumped out of the plane and was looking at the turned up leading edge of the tail section on his side. As I walked around the plane, he looked at me and with the clearest voice he said, "What are you doing, trying to kill me?"

We stood there in the snow, which was melting as fast as it hit, and marveled at our luck. No more than ten minutes went by and the snow stopped and the sun came out. I caught hold of the rolled up leading edge of the tail section and turned it back level. I said, "Bob, get in the plane."

Bob was still talking clear as a bell, when he said, "We can't fly with the tail like that." I explained that we flew for a few minutes after hitting the tree, so by rights it should be okay.

I taxied back to the same end of the runway and took off, having to clear power lines. In minutes we were back in the whiteout. I triggered the mike,

"Radio, Fairbanks, how is your weather?"

"We are in a whiteout. What is your number and location?"

"This is 22185 I just took off from Delta, and I am now returning."

91

Exposing Alaska

It had been a freak hole in the storm and it had now closed. I banked toward Donnelly Dome, picked up the 345 radial and started all over again.

Bob looked at me like a child and asked about the tree. I said, "Bob, the way I figure it, that tree was a loner and we cut it down to size, so we should have it made."

As the flag flipped, I began my slow count again. When I reached the same number, I banked 90 degrees to the left, dropped full flaps and cut throttle to 1500 rpm. Even I found it hard to believe for we sat back down, right on the end of the runway.

I looked at Bob and asked, "Now do we hitchhike back to camp or do you want to fly?" His voice was strong and clear when he said, "Hitchhike."

It was at least an hour before he began to stutter again. The flight was all Bob talked about for weeks, and I became a better pilot each time he stuttered it out. He undoubtedly had forgotten that God looks out for fools and drunkards and I was no drunkard.

Bob gave me a lift to Big Delta the next day and I flew to Fairbanks to have the damage checked. My home base was at Phillips Field and Mr. Bachner did the inspection. He had to order the new leading edge, but let me continue flying all summer. I flew up and down the glaciers, high in the mountains looking at sheep, often performing aerobatics with passengers.

It was August 28 before Mr. Bachner started the repairs. Instead of the damage being minor, it was major. All the main structure in the tail was cracked and could have broken off at any time. The repairs cost $3,077 and took twenty-eight days before I could fly again. This would not be my last close call while flying, but that's another story.

92

Exposing Alaska

We Were in the Air But We Couldn't Fly

Bill Griffin

You get scared sometimes and you don't realize you're scared until it's over. I've been unnerved a lot and I know because my mouth gets dry. I'd be flying along and everything would look good, but I would get an eerie feeling that something wasn't right, my mouth would get dry, really dry, and I would know I'm getting frightened. My dry mouth would come upon me like an uneasy warning, before I got into the situation where I was scared. But then a lot of things happen so fast that you don't have time to get scared because it's just an instantaneous thing. When you really have an emergency you'll panic for a second, but you've got to take control. Well, I had one emergency like that. Pete, the fellow I work with and I were out doing a survey. Every day we covered a lot of country, taking photographs and counting wolves.

Now, pilots know that flying a defective airplane or doing something you shouldn't be doing causes the majority of accidents. This airplane I was flying had an inoperative left brake. When I pushed the left brake that wheel didn't even slow down. When I pushed the right brake a lot would happen. That right wheel stopped and the plane would spin to the right. I knew it had a defective brake, but I didn't want to take the time to get it fixed. I took off because I wanted to keep on flying. So we were up there flying around for several hours counting wolves and we needed to make a pit stop. I spotted this strip, not too long, but it was the closest place to land, so we landed. Now beyond the strip was a huge lake, about five or six miles across and four hundred feet deep, and it was at about three thousand feet elevation. Before we landed, I could see that this lake was still about half covered with ice. We made a safe landing and I used the right brake sparingly as we rolled to a stop.

After a short break, we took off again down the runway. I noticed the plane wasn't developing power as it should have. Something was wrong; the power just wasn't there. I was getting close to the end and I didn't have enough power to take off. I started to panic but I couldn't, knowing I had to stay focused. So I touched the brakes and the plane spun to the right. I had no left brake and I knew I couldn't stop it. If I hit the right brake again, I'd go off the runway into a bunch of rocks. I'd probably land in those rocks and get killed. I couldn't do that. This all occurred to me in about a flash of a second. So I said, "Well, I'm just gonna keep on going." I gave it full throttle and maybe, just maybe, we would gain enough speed to get airborne. The next thing I knew, we were to the end of the runway where there was a ten-inch dirt berm. I couldn't stop. I hit that berm and bounced up in the air but I didn't have the speed or power to keep flying. I knew we were going down so I hollered back, "We're going in."

The right wing tip hit the water first, and it spun the plane around and the nose went in; we were under water. I tried to get out but I couldn't; the shoulder harness and safety belt was holding me snug. I fought off panic and concentrated on getting free. It felt like the airplane was going down, down, down, deeper, and I didn't want to drown strapped in my seat. I had to stop to think about what was holding me down. This only took a second but it seemed like an hour because I couldn't get out. I had to release my belt, then my shoulder harness, which is kind of hard to do under water, but I finally got loose. I was really scared, almost panicky under the water. I thought I was going to have to swim up, but the surface was right there. I gulped in a large quantity of air as I came up out of the water. Ten feet of the tail was all that was above water. As the air escaped, the plane was sinking. I hurriedly looked around to see if Pete had gotten out. I spotted him swimming for shore. Seeing that he got out was a huge relief to me. He was a strong swimmer. I had my leather boots on and warm clothes. It's very hard to swim wearing soggy wool clothes and boots in icy water. I tried to get up on the tail of the plane. I thought at first I'd climb up there and take my boots off. But it was like a muskrat trying to climb a greased pole. I slid off. There was no way I could get on the tail. So I said, "The hell with it." I pushed off and headed for shore. I had about 75 feet to go and I tried to swim but I couldn't, my boots were too heavy. So I thought, well, I'd stop and take my boots off. No, it was too cold. So I lay on my back and began kicking and paddling and finally I made it to the bank.

I was only in that 33-degree water three or four minutes, but that was really cold. For the minutes when I was under water and I couldn't get out of the airplane, I was just about panic-stricken. We both made it to shore, got up on the bank and watched our plane sink with all our gear in it. We had $800 worth of Fish and Game's camera equipment aboard; it all sank. That was a close call, too close.

Bone Chilling Cold

Sometimes You Wake Up,
Sometimes You Don't!

Bill Moberly

Outside, the temperature hovered at a cool twenty degrees as the softly falling snow added to the three-inch blanket already covering Fairbanks. My boss John and I were basking in the warmth of our electrical shop that day back in 1989 because work was slow. We sat there talking, the way men do, and naturally hunting came up. We swapped ideas about what to hunt for a few minutes when suddenly John said with enthusiasm, "Let's go deer hunting."

"Okay, great idea," I replied and we began discussing where and when to go.

First, we narrowed our choice to two popular deer hunting spots, Montague or Kodiak Island. "I hear that Kodiak Island has a larger bear population," I added, "and I'm not too fond of being bear bait." John agreed so we decided to go to Montague Island. A glance at the dusty wall calendar confirmed we only had ten days until Thanksgiving.

John suggested we invite our friend Ron to join us. It took Ron about two seconds to accept with a loud, "Yeah, man." Then, John called Harbor Air in Seward and made arrangements for us to be flown out a few days later. Harbor Air gave John all the information for the flight, including the maximum weight limit we were allowed.

Exposing Alaska

We needed to get our gear together so I started the grub and supplies list. Three men with big appetites on Montague Island for six days require a lot of groceries. With no 7-11 or any other stores around we knew if we didn't take it with us, we would have to do without. My list included all the grub, changes of clothes, sleeping bags, stove and lanterns, extra fuel, and everything else we thought we might need. The next day we began stacking the gear and supplies in the shop, and the pile just kept growing.

John stood six feet three inches in his stocking feet and Ron weighed in at a hefty 240 pounds. Mentally contemplating our load and knowing we had to come in under the maximum weight limit, we got the shop scales out and started weighing our gear and supplies. The scales confirmed we were indeed over our limits, and a discard pile began growing. We decided that our backup rifles not only weighed too much but that all of our rifles couldn't go wrong at once, so into the growing discard pile went the three extra rifles.

The day before our hunt, we loaded my pickup and headed south. We arrived in Seward late and rented a motel room. Next morning, the weather was foggy and rainy as we went out to Harbor Air to report in, and there we sat.

The weather began clearing about mid-day. Finally the sun broke through and we loaded up and departed. We enjoyed fair skies on our flight to Montague Island and thought everything looked favorable for a successful hunt. Our pilot landed smoothly and rolled to a stop on the beach near the cabin we had rented from the local Native association. With the tide coming in, we had to unload the plane in record time and then carry everything up to the cabin.

98

∾ *Exposing Alaska* ∾

It was a nice cabin with a wood stove and built-in bunks, and the outhouse was located about one hundred feet behind the cabin. We were camped on the bank of the Nelly Martin River in a six-mile-wide bowl. Mountains rose steeply on three sides of the bowl and the ocean gave us a great view in front.

We stowed our gear and with a couple hours of daylight left, we split up to take a stroll around and size up the deer and bear population. There was a lot of bear sign on the island, fresh footprints and droppings. Ron and I returned to the cabin before the sun set and we waited, but John didn't show up. Not knowing where he was, we lit a Coleman lantern as a beacon light.

It was a good hour after dark when we heard two shots fired. I fired a shot back, hoping it was John. It was and he was able to determine our direction and come on in. We listened attentively as John told of his past few hours. "I spotted a deer in a clearing and dropped it with one shot. I cleaned out the deer innards as darkness crept in around me and figured I'd better stash the carcass under a fallen tree, to keep it from the bears. By then the night was blacker than coal and I lost my way until I heard your return rifle shot."

"John," I said, "my dormant ulcer woke up worrying about you out there alone in the black of night in bear country. I would have been sick, had I known you were wallowing in fresh deer blood."

Three thick deer tenderloin steaks sizzling on that wood-fired stove created an aroma that put our hungry stomachs in attack mode. I removed the frying pan and placed a steak with still a trace of red in the center on each paper plate. Those steaks and a half loaf of buttered bread was our

99

dinner. After the growling had ceased in our contented midsections, we crawled into our warm sleeping bags.

Rays of early morning sunlight peeking through the window woke us. In preparation for an eventful day of hunting, I prepared a hearty breakfast of sourdough cakes and bacon. Then we hiked up to where John had stashed the deer and his pack board. A bear had beat us to it; the pack board was ripped to shreds. The bear had pulled the deer out from under the tree and had taken big bites out of most of the deer, and what it didn't chew on, it had rolled around in the dirt. A sinister awareness of bears overshadowed us.

I suggested we stay together now for our own safety as we continued walking the trails. Our plan was to get more familiar with the terrain before going back to the cabin for lunch. A bear had ruined the first deer, but we knew there would be more. We hiked up the side of the mountains that surrounded that oval-shaped bowl of a valley. It started raining on us and a strong wind came up. First thing we knew, we were in a full-fledged nasty storm. Fifty-to-seventy-mile-an-hour winds stabbed at our faces and the rain poured down. We knew the cabin wasn't too far from the beach so we headed toward what we thought was surf breaking. But no matter which direction we walked, it sounded like the surf crashing on the beach. After walking around out there for awhile with no ocean in sight we determined it wasn't the surf we were hearing. It was the wind whistling through the treetops. Then the dense fog rolled in so thick we couldn't see in front of us. We became completely disoriented and had no idea what direction the beach was. We'd carelessly left our compasses and maps back at the cabin. Now being clueless, we wandered around out there for hours, stumbling, falling, and cussing.

∽ *Exposing Alaska* ∽

Adding to our dilemma, it gets dark early in November on Montague Island. John and I had little Duracell pocket flashlights and thought, "We can't go too far wrong, we'll just keep walking." We were hungry, cold, wet, and miserable. The chill factor increased because we were soaked and had a strong wind at our back. We followed a stream that we could practically step across. It wound through the valley and we wandered with it. We waded the stream in places and crossed over fallen trees in others. The terrain grew worse as we trudged in deep mud and crawled and stumbled through thick underbrush. Sometimes we dragged our rifles behind us; other spots were impassable.

The night grew blacker, the fog denser, and the wind howled like a pack of wolves. We got wetter, colder, and more lost. The stream was chest-deep the last time we waded it. Now as midnight approached, we were chilled to the bone and then, both our little Duracell flashlight batteries died.

Each of us well knew the seriousness of our perils, lost on an island, soaking wet and in real danger of freezing to death. It wasn't quite so bad as long as we kept moving. But when we stopped to rest, our core temperature dropped and all three of us began to shiver violently, a sure sign of hypothermia. We kept falling and getting up and staggering around as if we had spent the day inhaling whisky. We knew it was foolish to keep stumbling around in the dark and possibly breaking a leg. So for protection from the penetrating cold and the fierce wind-driven rain we decided to crawl under a big tree stump on the bank of the river. We tried cutting the roots off the stump with our hunting knives, but the wood was green and wet. Trying to get a fire started proved no more successful than trying to ignite a

101

drenched newspaper. I had worked at Prudhoe Bay and had experienced high winds and one-hundred-degrees-below-zero chill factors before, but I had never felt this cold or miserable.

Ron was shaking and talking incoherently through chattering teeth. I heard him muttering something about, "Which one of you guys is responsible for us being out here freezing? I don't want to die like this." Neither John nor I answered him.

I got the bright idea to pull the shells out of my ammunition and used the gunpowder to start a fire. That didn't work; the gunpowder just flared up and vanished. I looked at my Winchester 300 H & H Magnum, my pride and joy, a rifle I had bought back in 1960. I thought, "Well, there's some dry kindling on that stock." I broke the stock off and we got a little fire going with the splintered rifle stock. We added some of the smaller roots from the stump we were under. We crowded tightly around that little-bitsy blaze; and it brought our spirits up a bit, but the stock burnt up in five minutes.

I tried to burn my wool stocking cap, but it was too wet. The cold crept deeper into our bodies, and we became more miserable as our chances of survival grew bleaker.

John had a Belgium Browning that he thought quite a lot of. He said, "Well, we got a couple more rifle stocks; maybe we'll survive till daylight." So we broke the stock off the Belgium Browning and got another small blaze going. Last came Ron's Remington 30.06. We broke it off and got another tiny fire going that warmed us a little as we hunkered down and cuddled up close.

Time seemed to stand still and my thoughts drifted back home to my wife Marge. "Oh God, I wish I were back there with her right now in our warm comfy bed."

∽ *Exposing Alaska* ∽

After an eternity, the blackness of night faded and traces of light appeared. My body felt numb from the longest and coldest night I've ever spent. I creaked as I crawled out from underneath the roots of that big tree and onto the bank of the river. In the wee bit of morning light I could see that the fog had lifted a few feet off the ground. Squinting, I could make out a well-used game and people trail on the bank of the river. I followed the trail about eighty yards and came to a Forest Service cabin with smoke curling out from the stovepipe. I knocked hard and three startled GI's, hunters from Ft. Richardson, let me in. These nice guys had a big warm fire in their glowing barrel stove and they invited us in to warm up. I hurried back and got John and Ron, and we returned to their cabin and began hugging their stove. A steaming cup of hot coffee helped melt my frozen insides as my outer body soaked up the warmth radiating from that stove. It felt heavenly! The GI's stood with open mouth and in shock as I unfolded the story of our previous night, and the dilemma that caused three hunters to break the stocks off our rifles.

After we'd warmed up a few hours, we thanked our benefactors and left, carrying our stockless rifles. The wind continued blowing and the rain poured down like from a full open faucet in the sky. In the faint daylight we could tell where we were. We hurred back to our cabin and built a big, roaring fire in the stove and made some hot soup, and then we crawled into our sleeping bags to make up for the sleep we had lost.

We woke up late on the third morning of our hunt to the stark realization that we didn't have a useable gun to do any deer hunting. Worse yet, we were defenseless on an island that was alive with bears. Our spare rifles were on

103

top of the discard pile back in the shop. I felt safer running all the way to the outhouse.

We three big deer hunters spent the next five days in the cabin, sleeping, eating, playing pinochle, and pondering why three grown men had gotten in a situation where we could have frozen and died. We were very grateful to be alive and warm again. The weather cleared and Harbor Air flew in to get us on schedule. We did feel kind of sheepish walking out to that airplane with our three stockless rifles and no deer.

We drove back to Fairbanks quietly and arrived home at 4:00 AM Thanksgiving morning. I tried to be quiet while taking a much-needed shower but Marge woke up. I held her hand tight while I told her of the perils we'd been through. I told her how much I loved her and missed her before crawling into our warm cozy bed, the very same bed I missed so desperately while freezing on Montague Island.

John, Ron, and I had a lot to be thankful for that Thanksgiving. After making all the mistakes we had made—we survived. We didn't have any venison, but nobody lost fingers or toes to frostbite either. We three will never forget our little escapade and we thanked the Lord that we were alive after a night out like that. Because, when you are so cold that your extremities turn numb, it's real easy to just lie down, give up, and go to sleep. Sometimes you wake up, sometimes you don't!

I Don't Like Being Cold

Bill Griffin

Every winter I cuss Alaska. I record on my little tape recorder, "Next winter I'm going to leave this frozen state." Below zero temperatures in Alaska are just too dammed cold. I'll never stay here another winter. I've made that statement and meant it for near half of my 70 years.

I'll go back to western Oregon where the climate's mild. In the wintertime the grass is still green and the trees are giants; I like that. Tennessee is where I came from originally but I left there and moved to Denver, Colorado, for awhile. Next I moved to Washington, then to Utah, where I went to college for four years, then back to Oregon. I like Oregon.

I first came north to Alaska thirty-four years ago on a two-week sheep hunt. After the most interesting and successful hunts of my life, I was hooked. The two weeks went by too fast. I returned to Oregon to my suit-and-tie job and to the structured suburban life style. I wasn't content any more. Something about Alaska was drawing me back: the freedom, the wide-open spaces, the uninhibited lifestyles, the hunting and fishing, the mountains. I just loved it all. Living in the wilderness attracts a certain type of person, and I am that type. I love being out in the wilds where there is an element of danger.

∽ *Exposing Alaska* ∽

So, filled with anticipation and excitement, I applied for a job with the Fish and Game Department in Alaska. I'm a biologist and a pilot and I was overjoyed when they hired me. This was my dream come true.

I love flying to places most Alaskans never get to see. I enjoy the freedom of bush flight in a two-seated plane, darting across the heavens like an eagle in flight. I had no interest in flying the huge commercial jets. That straight as an arrow, point A to Point B, cross-country flying was not for me.

Now, bush flying was part of my job. I took surveys of the various types wildlife. My job was counting moose, caribou, sheep, seals, grizzly bears, and polar bears from the air. Once I conducted a survey, thirty days just counting falcon nests on the Seward Peninsula, finding bird nests up in the cliffs with an airplane. That was the best job I ever had, working was like being on vacation. Every day was exciting, flying low and counting birds and animals.

I remained with the Fish and Game Department until I retired. Since then I've been a flight instructor, flight examiner, and an air taxi pilot, and I still love to fly; I'm thankful I'm still able to soar across the skies with the eagles. Alaska is pilot country. If you've got an airplane it makes life double interesting to live up here.

When I take a trip outside to Oregon, I realize how different our way of life is. Alaska is a contrasting lifestyle and so open, I miss it. That's why I've spent the last thirty-five years in the north; that's the reason I'm still here.

The weather is always fine for the fellow who is appropriately dressed.

I've Been in Seventy-eight Degrees Below Zero

Bill Griffin

I landed my super cub in Bettles one night and went into the lodge. The surprised owner greeted me with, "Do you know it's seventy-eight degrees below zero outside?" I said, "No-ooo." He said, "I've been talking to Prospect Creek here on the radio, it's seventy-eight below zero."

I couldn't believe it; I've never heard of anything that cold. So I threw on my parka, pulled the hood over my head, put on my mitts, and walked outside again. I went out to experience the sounds and feel the bite of this record cold. I pulled my hood down and listened: quiet, absolute stillness was all I heard. My ears immediately began to ache, calling for the return of my hood. On opening my mouth to exhale, my breath froze, the frost clung to my beard and mustache as if it were glued on. I inhaled a short breath of that minus seventy-eight-degree air and felt the resistance in my windpipe. My lungs were gasping for more oxygen while the rest of my body was trying to shut out this cold intrusion. I removed my mitts in order to touch and feel the air, hoping to gain endurance for cold of this magnitude. In the time it took to pry the ice off my eyelashes, my fingers reacted to this frigid exposure in three ways. First they tingled and hurt, then they stiffened up and lost feeling, then I noticed the tips of my fingers turning white. So much for the feel of cold. I struggled to put my mitts back on.

I walked around out there for ten minutes, just to really know what this record-breaking low temperature felt like. It was a real thrill for me. When people ask me why I stay in the north, I tell them of the freedom, the wildlife, and the mountains, all which intrigue me and hold me here. I tell them there's no place like Alaska. I've been in seventy-eight degrees below zero, and I'm kind of proud of that.

Twenty-four Hours in Fifty Below Zero

Tasha Custer

We left our home in North Pole early in the morning of January 8, 1995. Even though the temperature was fifty below zero, Mom agreed to drive my four-month-old baby brother and me to my uncle's birthday dinner in Anchorage. It was definitely the worst traveling weather.

The first 250 miles were okay. Mom drove and I cuddled my brother. Suddenly our car quit on a deserted stretch of highway about one hundred miles from Anchorage. Mom didn't know what to do, but finally she decided that the solution would be to stand outside the vehicle and wait for someone to come by and give us a ride to the nearest phone. I sat in the car with the baby. He began to cry. I worried and wondered what was going to happen to us. I could tell Mom was very upset about our situation and the baby's welfare. After twenty minutes of standing outside in the cold shivering, Mom crept back into the car with tears frozen on her face. With no sign of anyone coming we sat there silently in our cold car as the lonely hours slowly passed.

In our effort to keep ourselves warm we huddled together in the backseat with my baby brother between us. Our warm clothes and the one blanket we had brought were our only protection as the temperature inside the car matched the outside deep freeze. We spent the night, shivering together. It was the longest, coldest, most miserable

Exposing Alaska

night I could ever imagine. When dawn's light began peeking through the icy windows, Mom got out of the car again to resume her wait for our rescuers to come. Finally a car did come along and although the people couldn't give us a ride for lack of space, they allowed my mom to use their cellular phone so she could call my dad to let him know what had happened. As that car drove out of sight, I got a pain of uneasiness in my stomach. I feared we would be here another whole day, and I didn't think my little brother would survive without milk or anything to eat or drink.

Just as my mom had climbed back into the car and shut the door we again heard the welcome sound of an oncoming vehicle. My mom stepped out of the car again and wildly began flagging them down. A big blue van rolled up to us and the drivers side window was rolled down to expose a friendly face. The kind man helped us pack our things into the back of his van and gave us a ride to the nearest roadhouse where we were able to warm up and eat while we waited for my dad to come. Alaska's harsh winter had held us at its mercy but thanks to that kind man and my dad, we were back home that night, safe and warm in our own beds.

The record snowfall in a season was 974.5 inches at Thompson Pass, near Valdez, during the winter of 1952–53. The record snowfall in twenty-four hours was sixty-two inches, at Thompson Pass in December 1955.

Denali National Park is slightly larger than Massachusetts. Inside its boundaries is Mount McKinley, the highest mountain on the North American continent. Its south peak is 20,320 feet high.

The Indestructible Moose

Paul Elbert

Getting caught in one of nature's cruelest winter traps most often ends in death to man or beast. Crossing a river on thin ice, in below-zero weather, is a treacherous walk right into an icy-cold enclosure. Men die in minutes; that a moose can withstand such an ordeal was a revelation to me. I will call this moose Jake.

Jake was a twenty-two-month-old friendly male moose who grew up in the vicinity of Fairbanks, Alaska. For months he had been wandering un-harassed by dogs or people through the suburbs of Hamilton acres and Island Homes. All was well until one day Jake decided to cross the Chena River and walk into the very heart of town. Forty howling huskies in a kennel nearby did not deter him. What Jake did not know was that thin ice lay hidden under a thick blanket of snow.

Mid river, the ice broke. Jake's feet just touched the rocky bottom in four feet of bone-chilling water. By jumping and treading he could keep afloat, but there wasn't a chance of his climbing out. He couldn't raise his feet high enough to get above the ice. The temperature stood at twenty degrees below zero. This ice kept him prisoner, ensnared like the bars of a jail cell. How much longer could he live?

George Kennedy, who lived across the river, saw the moose in distress and called the police for help. The police

⌒ *Exposing Alaska* ⌒

and firemen arrived and managed to get a rope around Jake's neck and were able to winch him out. George was helping when he fell through thin ice, almost losing his life after helping to save the moose.

Jake had been in the ice-cold water for about one hour. That the exposure did not cause the blood to congeal and stop his heart is a revelation of built-in insulation and hardiness. Jake lay in a state of shock and exhaustion for about an hour. Firemen covered his body with a blanket and an Eskimo observer heaped snow over his feet and legs.

Finally, Jake got up and struggled off to some bushes to feed. The blanket fell off and subzero air hit his wet body. He became so encrusted in ice that he could hardly move or even raise his head. Jake browsed on a willow bush for half an hour and then lay down in the soft snow. After a rest he got up and browsed again. But he was visibly weak and looked done in. I doubted that he would survive the night.

This tragedy happened on March 14, and our temperatures were running below normal. The thermometer registered thirty-five degrees below zero that night. Wet to the skin, encased in ice and snow, Jake would certainly freeze to death.

The next morning, I returned to the scene. I fully expected to find Jake frozen to death. But no, Jake had crossed Clay Street to another vacant lot and was browsing, head a little higher than the day before. I took several pictures and was fascinated by his heroic feat. I just spent the day observing him.

By noon, the sun's warming rays had brought the temperature up to zero. Jake began to steam from stem to stern.

112

Every so often, he shook himself and ice fragments flew in all directions. By three o'clock, his ice cast was nearly gone.

I returned again the morning of the sixteenth. This moose's survival fascinated me. I wanted to be near it, to see for myself how it recovered and behaved. Jake was browsing quite normally and held his head much higher. He had browsed for three days on the low branches of a dozen willow bushes on two vacant lots. Now he was able to lift his head up for higher branches.

Moose are insulated to withstand our blizzards and subzero weather and a little bit more. Some moose die each winter because disease, parasites, or predators have weakened them. Otherwise, healthy moose like Jake, if they learn to stay away from cars, railroads, and hunters, are well nigh indestructible.

The "Trail of '42"—the Alcan, which was constructed in 1942–43, has progressed from a trail to the paved Alaska Highway of today. You'll enjoy the all-Alaska sport of dodging potholes and avoiding dips, and frost heaves on most of our roads. Be advised however: they are strategically placed so you can't miss them all.

Surviving Alaska's Relentless Waterways

Surviving the Chatanika

Chad Dietz

My waders filled up with water and I went under. The strong current washed me downstream where the river turned into a torrent. I was just trying to catch some air and breathe.

My parents moved to Alaska in 1984 and I followed them. A nineteen-year-old cheechako from North Dakota, I wanted to experience all aspects of life on the Last Frontier. My dad had told me all about the great fishing in Alaska and the record fifteen-inch grayling he had caught. Dad elaborated on the superior taste and how grayling melted in his mouth after being seasoned and timely cooked over an open campfire. I couldn't wait to try my own skills at fishing and was anxious to try to break Dad's fifteen-inch record by pulling in a granddaddy grayling. Unfortunately, with all this unbridled enthusiasm, my first trip was almost my last.

One of my new friends in Fairbanks is an Alaska Native named Abear. In talking with him, I discovered we had a lot in common. We were both outdoor kind of guys and we both loved to fish. After much pleading on my part intermingled with some unmanly begging, he invited me along to his secret fishing hole on the following Saturday.

All week long I brimmed with anticipation, thinking about the fish that awaited us and of the thrill of catching them. Finally Saturday morning arrived. We left Fairbanks in my old pickup with Abear directing the way. We followed the old road up to the top of Murphy Dome and then continued on a narrow dirt trail down the back side to the Chatanika River. I parked beside a birch tree, bent from the winter snow load; we sat on the tailgate to remove our shoes and slid into our chest waders. Fifteen minutes of hiking upriver brought us to the spot. Abear said, "It's across the river beneath that undercut bank just off the main current. That's my spot; I've caught some beauties out of that hole."

Excited, we both baited our hooks and walked out in the current to catch our limit of prize-winning grayling. A half-hour went by and nothing, not one bite. Cast after cast, including numerous bait changes, mepps, flies, and spinners from the well-supplied tackle box Abear had lent me, and not even one encouraging nibble.

I became disillusioned and started doubting Abear's stories. The only response I could get out of Abear was, "Oh, you cheechako are all the same, in a hurry. Ya gotta learn to be patient; there's big grayling hiding in there."

I continued to grow more discouraged since we weren't having any luck at all. Finally I told Abear, "I think you caught all the grayling out of this hole; I'm going to look for another spot." So I took off down river looking for a productive fishing hole.

I came to a place where the river branched out and was running in three separate channels. The main channel where I wanted to be was the far one. I crossed the first two channels with no problem, since they were just knee deep.

115

∾ *Exposing Alaska* ∾

But once I was out in the third channel things got weird; it was deeper and the current stronger. The water was pushing hard against my legs and rising higher toward my chest. The force was overpowering, and then I lost my balance and fell in. Letting go of the fishing pole and tackle box was an automatic response. My waders quickly filled with water and I went under. I was being washed down stream where the river turned into a torrent. Now I was just trying to breathe and stay alive. I manage to turn my face up and grab a gulp of air occasionally. My waders were full of ice cold water, numbing my legs and weighing me down like lead shoes. I tried to remove them but I couldn't get them down past my knee. At this point I couldn't even touch bottom any more. Then I got caught in a whirlpool, going around and around. I was choking, sputtering, and in desperate need of air. I knew that I couldn't last much longer and I prayed, "God, please don't let me drown out here; please just give me something to hold on to."

By now I was floating by a beaver dam and that's when God answered my prayer. A branch reached out and hit me in the hand. I grabbed it and pulled myself toward the beaver dam where my feet touched gravel. Then the branch pulled loose, but not before I got my footing. The water was calmer now below the beaver dam and I was able make it up on the bank. I collapsed and lay there on the bank spitting water and coughing. A sigh of relief and a thank you, Jesus, prayer audibly escaped with my panting breath.

It was forty-five minutes before a very concerned Abear sloshed downstream to where I lay. After seeing that I was all right, he immediately reacted in anger because I lost his well endowed-tackle box.

116

∽ *Exposing Alaska* ∽

I was just thankful my prayer had been answered and happy to still be alive and out of the ice-cold water. I told Abear, "I will be more than happy to buy you a new tackle box filled with the newest and finest fishing lures." At that moment, just being alive took on a whole new meaning for me.

The longest day at Prudhoe Bay is sixty-three days, twenty-three hours, forty minutes—from May 20 to July 22.

The longest night at Prudhoe Bay is fifty-four days, twenty-two hours, fifty-one minutes—from November 24 to January 18.

The shortest day at Prudhoe Bay is one hour, three minutes on November 24.

The shortest night at Prudhoe Bay is twenty-six minutes on May 19 and 20.

The highest temperature recorded at Prudhoe was eighty-three degrees F. on June 21, 1989.

The lowest temperature at Prudhoe, minus sixty-two degrees F., was on January 27, 1989.

The highest wind speed at Prudhoe, 109 MPH on February 25, 1989.

The official low chill factor was recorded February 28, 1989. It was minus fifty-four degrees with thirty-six-mile-per-hour winds, which gave us a chill factor of minus 135 degrees F.

117

A Riverboat Nightmare

Don Elbert

It all began safe enough that spring afternoon back in 1969, with an on-the-spot purchase from the sporting section of our Fairbanks Pay-N-Pack store. My adventurous nine-year-old son, Kenny, kept pulling me toward the rowboats on display, pleading, "Dad, let's buy this one; we can row it all over Ballaine Lake. Please Dad, it will be so fun."

The item he pointed to was a fourteen-foot lightweight aluminum rowboat with two seats and wooden oars. I looked it over and it appeared sturdy enough. I lifted one end to check the weight and determined it was light enough to carry. Another glance at Kenny's gleeful face, brimming with anticipation, combined with the chorus of pleading from our four other children pretty well cinched that decision. So for $249.99 we bought our first boat, and my love affair with river running slowly emerged.

Together my joyful family helped carry it out and we loaded it on top of our Ford station wagon. After strapping it down, we headed for nearby Ballaine Lake. The children delighted in rowing each other, as well as mother and me around the lake. In the days and months that followed, we spent many of our summer hours rowing, frolicking, and picnicking on or beside the water.

∽ *Exposing Alaska* ∽

The following summer, I bought a used twenty-foot flat-bottomed aluminum riverboat with a forty-horsepower Evinrude from a neighbor. Having a motor sure beat rowing. Also, it opened up a whole new world of adventures as we traveled into the wild side of Alaska by its rivers. The freedom and joy I experienced while boating over the next eleven years grew into a passion for exploring the river highways into the unpopulated regions of Alaska. Plus, I was able to reach the richer fishing areas and catch my limit of graylings. My river travels usually climaxed each year when with one shot, I'd shoot enough moose roasts to feed my family of seven all winter.

In my continuing search for the ultimate in trouble-free river transportation over those same eleven years, more boats came and went in a series of boat upgrades The next one was a well-worn twenty-eight-foot wooden craft that I had to refiberglass the bottom to keep it afloat. Its fifty-horsepower outboard Mercury with prop pushed it upstream at nearly twenty-five mph. Then a good buy came along on a twenty-four-foot wooden inboard jet boat with a 283 Chevrolet engine that had thrown a rod. After a week of rebuilding the engine, I fired it up, and we again enjoyed the rivers leading into the spacious wilderness portions of Alaska. These larger boats were great for carrying the whole family, more gear, barrels of gas, and a moose. However wooden boats are really heavy when trying to pry them off the many submerged sandbars lurking just beneath the water. Having had that experience several times, I reverted back to a twenty-foot lightweight aluminum model pushed by a forty-horse Evinrude outboard with a new jet unit. That outboard worked well for many years on our shallow rivers.

The entire family enjoyed the longer trips we took together on the Tanana River to Old Minto and on to Manley Hot Springs. One trip took us up the Nenana River into Mt. McKinley Park. Other rivers we explored included the Salchaket Slough, Dry Creek, and Bear Creek, 110 miles up the Salcha River looking for moose, then up five-mile Clear Creek, on down to Big Delta, and up the Goodpasture River as well as Harding and Birch Lakes. However, these excursions also included more than my share of river mishaps and major boating malfunctions. Things like rocks punching holes in the boat, being high-centered on sandbars, seized-up engines, and running out of gas were just a few of them.

One summer, while out scouting for a closer possible moose hunting area with my hunting partner Bob, I made the wrong choice between two channels. Forty-five miles up the Salcha River I ran my thin-bottomed aluminum boat over numerous, submerged rocks hiding in less than two inches of flowing water. One knife-sharp rock peeled the metal back like a can opener. The Salcha poured in and I raced for the closest bank. After jumping out we pulled the boat as far up on the bank as we could, which was not far because it was half full of water by then. Bob quickly began bailing while I pulled up floorboards, searching for the hole somewhere in the back of the boat. Finding it, I bent the thin strip of aluminum back and Bob took off his boot, removed a wool sock, and jammed it in the remaining hole. We again bailed out water before pushing off from the bank and raced the hour and a half downriver to the bridge and my trailer.

I worked for the U.S. Army as lead mechanic in the Heavy Equipment Shop on Fort Wainwright. Bob, a young sergeant, had rotated into our shop sixteen months earlier.

∾ *Exposing Alaska* ∾

Bob was a wholesome, lanky, corn-fed farm boy with powerful blue eyes and a short military crewcut who enjoyed the outdoors. I soon learned that Bob's great loves were hunting and fishing. Being a single fellow in his early thirties, he filled his off-duty hours doing just that, hunting and fishing. Last year we hunted together and Bob shot the only moose on that trip. We had made several trips together in my riverboat. I enjoyed Bob's humor and appreciated his help when we had a problem. We had more than our share of those.

The next spring, I purchased a dreamboat: a sturdier, sleek and shiny, Valco inboard jet craft with a $^3/_{16}$-inch thick bottom. This eighteen-foot beauty with a six-foot wide beam was powered with a new inboard four-cylinder Chevrolet engine. Inside my garage, a meager pile of camping gear, supplies, and groceries kept growing taller all summer long. A bountiful moose population was reported on the Nowitna River, and we planned on being there and shooting one.

We know that the last ten days of the hunting season, when moose are rutting and looking for a mate, we'd have a better chance of spotting one. We were filled with high expectations that morning, September 10, 1978. Bob's wide grin spoke of success as I backed my loaded trailer down into the Chena River near Pike's Landing. After loading the boat in the thirty-six-degree coolness, we departed Fairbanks via the Chena, which flows into the Tanana River, and headed for the Yukon River.

We hoped all our old boat problems were behind us now as my new sleek jet boat skimmed over the water and down the river. By midafternoon we had covered 225 miles and reached the mouth of the Tanana, where it spreads out

∾ *Exposing Alaska* ∾

before flowing into the Yukon River. A loaded barge was mired down on a submerged sandbar. There was nothing we could do to help. It took all the strength we could muster just to pry my boat off a sandbar. So, we crossed the Yukon River to the village of Tanana where I had planned to top off our gas tank.

On arriving in the village, we discovered that their only gas station was closed. The large blue crayoned words on the cardboard sign tacked to the front door read "OUT of GAS." The whole town was waiting for a delivery from the barge stuck on the sandbar across the river.

As a last resort, I hoped to buy gas from any of the town folks if they had some to spare. Bob stayed to keep an eye on my boat while I walked down the dusty main street of what appeared to be a nearly deserted village. The noise level from the local sled-dog population nearly deafened me. Spotting an older Native lady cutting fish beside her cabin, I asked her if she knew of anyone who had gas to sell. She pointed to another low, sod-covered log cabin down the street and replied, "Talk to John."

I thanked her and with a glimpse of hope, headed straight for that cabin and knocked, even though the door was open. A muffled voice spoke: "come on in." I entered, crouching under the low ceiling into a dark and dingy room. Straining my eyes to adjust to the lack of light I noticed several men sitting on the floor drinking. The open whiskey bottles being passed between them should have given me a clue of what was coming. They all stared at me as I told them I was looking for John and wanted to buy gas and would pay double the cost. A brown-toothed, scruffy fellow, I guess it was John, held onto the fellow next to him for support and asked in a sputtering of words, "Where ya from and where ya think ya going?"

122

∾ *Exposing Alaska* ∾

I replied, "My friend and I are down here from Fairbanks, and we're going moose hunting up the Nowitna River."

I watched the tension mount as he clinched his fist, gritted his rotten teeth and stammered, "You damn city hunters come down here killing all our moose. Why don't you just get the hell out of Tanana?"

I looked toward the door, intending to do just that—get the heck out of there. Then another fellow got up staggering and led me outside. Excusing his friend John's actions, he confirmed that the whole town was indeed out of gas.

So Bob and I left Tanana without extra gas. I figured if we eliminated running up and down the Nowitna River looking for moose we should make it back to Tanana on the return trip. The station would surely have gas by then.

Hours later and 308 miles down river from Fairbanks, I spotted the inlet and turned from the quarter-mile-wide Yukon into the narrow mouth of the Nowitna River. The temperature held steady at a cool thirty-six degrees and the thick mist turned into a light rain.

My jet boat continued to perform well, skimming over the sometimes narrow and often shallow water of the snake-like Nowitna. The next two and a half-hours brought us eighty-five miles up from the mouth as we slid sideways around another sharp bend in the Nowitna. Suddenly, I heard a horrible sound, one I dreaded hearing, the metal-to-metal rattling and banging noise an engine makes without lubrication. I looked back and witnessed a steady stream of black engine oil spurting out. Immediately I killed the engine and we drifted to the closest bank. The copper line leading to the oil gauge had broken due to the vibration, allowing the engine's oil to pump out. Most of the four

123

∽ *Exposing Alaska* ∽

quarts capacity of essential engine lubricant now covered much of our gear and the several inches of accumulated rainwater in the back of the boat.

We were out of commission, dead in the water and 393 river miles from home. All of a sudden our well-planned hunting extravaganza had turned into a plight of survival. We hadn't passed another boat or seen anyone in the last two hours. The drizzle had turned to a steady rain; we were wet, cold, tired, miserable and terribly depressed.

Bob tied the boat to a tree root; we unloaded the tent and a few supplies and selected a spot to camp back in some small trees just up and off the river. After cutting a birch pole for a center support, we set up the tent in the mud and moved inside. We were a couple of sad hunters as we dropped into our sleeping bags that night.

I didn't sleep; my eyes stared into the blackness. As winter crept in closer each day now, the majority of the smarter hunters had departed. The sounds of the rain beating down on our tent interrupted my blurred thoughts of slow floating 115 miles down river to Ruby for help. I grew more miserable as I envisioned a future hunter finding our bare bones in the spring.

The next morning we discovered that our tent leaked in several places and the bottom half of my sleeping bag was wet. It had quit raining temporarily, so I went down to check on the damage to my boat. The $1/8$-inch brass fitting that the copper oil line screws into had broken off flush with the block. I took apart a small pair of ignition pliers to use as a tool that would fit snugly inside the brass threads. Then I prayed it would not strip out as I gently began to turn. It broke free and I was able to extract the brass threads. We sorted through the toolbox and Bob found a eighth-inch pipe plug that I screwed into the hole. Now we just needed more oil; I had only brought one extra quart and needed a total of four.

By mid-afternoon, the welcome engine whine of another jet boat grew louder, and it was the most pleasant sound we could have heard. Two wary hunters in Army jackets and wielding rifles and the boat driver came around the bend in the river. We jumped up and down, waving our coats to flag them down. They stopped, and I told them of our dilemma. Then I asked the big question, "Can you spare any engine oil?"

The driver, a tall man with wind-blown hair and a friendly face, introduced himself. "Hi, I'm Vern McConnell. My old Chevy V-8 engine burns oil so I brought along ten extra quarts." And being a true Alaskan and seeing our plight he gave me the three quarts of oil I needed. I did thank him but he wouldn't let me pay for the oil. We bid each other happy hunting and they pushed off and returned down river looking for a moose. I poured in the oil and turned the key; the engine cranked over. With a pull of the choke it started but rattled and clanked until the lubricant reached the dry spots. I didn't have a working oil gauge anymore but I could tell we had oil pressure. The quiet hum of the engine was my sure-fire stress reliever. I felt vibrant again as my major boat concerns vanished.

It started to rain again and it continued all week. We didn't use the boat for hunting, opting instead to save enough gas to get back to Tanana. Most of our time was spent in the tent trying to keep warm and dry and bemoaning the rainy conditions. Light from my Coleman lantern enabled us to read the four Western pocket books Bob had brought along—twice. For nourishment we ate enough peanut

125

Exposing Alaska

butter and cheese sandwiches to plug up a horse. Then we washed it down with cup after cup of strong, Nowitna River coffee. Considerable time was spent wringing out wet clothing and moving our cots and gear to avoid the leaks. The air we breathed was saturated with the pungent stench from stinky wool socks and muddy wet clothes hanging to dry over the Coleman stove. Our own unwashed bodies added significantly to the ripe aroma. The only good side of this situation was that the awful smell inside and the constant rain outside kept the insects away.

We talked a lot to break the monotony. Bob reminded me of a trip we took a month earlier, the maiden voyage in my inboard jet boat. My wife, Elaine, and I had invited Bob and his girlfriend Kris along. We were out scouting again, about an hour up the Nenana River this time, when again I went high centered on a gravel bar. All four of us stepped out of the boat into about two inches of water and tried to push the boat back in the river. We couldn't budge it and quickly realized that the stuff that makes boats stronger also makes them heavier. Bob volunteered to wade to the bank and cut some pry poles with our camp saw. Kris chirped, "Wait, I'll help you." They waded through thigh-high water, hand in hand, the twenty feet to shore. Bob found several dry poles high up on the bank and Kris helped trim them. They returned with four pry poles. We wedged them under the boat, then lifted in unison. We could only move the boat an inch or two each try. Laboring for an hour and forty minutes, we pried and pushed in an all-out effort to get off that gravel bar and float the boat again. Kris kept our spirits up by kidding Bob. She said "Come on Bob, get your back into it." Then she reached over and tickled him. Kris laughed the loudest when Bob's pry pole broke and he toppled over in the water. Then in her comforting way she sat down beside him in three inches of water and gave him a hug. Elaine was more serious, focusing on getting off the rocks and back home to our kids. In one of the last big prying efforts we heard a rip. All eyes turned to Kris. She had split her britches, but even that didn't dampen her spirit.

From his cot on one side of our wet tent, Bob told me that he was so impressed with Kris that day, her caring and sharing and willingness to help ease the boat back in the water. That's when he made the decision to ask her to marry him. Now Bob was thinking of Kris and he lamented, "I wish I was back home with Kris right now, cuddled up warm and contented."

I replied, "Me too. I would much rather be home with Elaine and our kids, sleeping in my own dry bed instead of this wet, clammy sleeping bag. Who the heck's idea was it anyway to make this trip so far from home?" Bob looked at me with that goosey grin of his and we both broke out laughing.

Late on the third night of our wet adventure, Bob had climbed in his sleeping bag and I had slipped out to brush the peanut butter off my teeth in the downpour. A guttural grunting noise every few seconds startled me from across the river. These grunts were getting closer and louder. I knew a bull moose was approaching. I sloshed back to the tent, grabbed my rifle, and whispered to Bob, "Get up, a moose is coming." I hurried back out in the rain and peered through the willows. With just enough twilight to see movement I first noticed something in the river. Moments later I recognized moose antlers coming up out of the water. The antlers were mounted on the head of a huge moose.

∽ *Exposing Alaska* ∽

The grunts were his cow calls; he was looking for love. He climbed up the bank seventy-five feet from me. After wiping the dew from my scope, I raised my 30.06 and placed one shot just behind the front shoulder and he went down. That's the way my Daddy taught me to shoot. Bob, bare footed and wearing his long-johns, came rushing out, high stepping in an inch of mud with his rifle drawn, just in time to see the moose fall. Bob's first words were, "Man, that is one huge hunk of moose. Good shot, Don."

The next several hours were spent out in the rain, gutting the moose by the light of a kerosene lantern. We saved the heart and liver in a pan then covered the carcass with Visqueen for the night. The next day we cut birch poles and made a hanging rack eight feet up off the ground. We quartered the moose, then removed the rib cage, and with the help of a come-along, hoisted each piece up on the rack to air-dry and season. We covered the meat with Visqueen to keep the rain off. The antlers measured sixty-three inches across; they were keepers. The rain continued to dump on us. We should have started for home but Bob still hoped to shoot a moose of his own, and I was hoping the rain would stop.

Two days later, just past midnight, Bob shook me awake. His voice was full of alarm. "There's a bear out there getting our moose." We both jumped up, grabbing our raincoats and rifles. Bare-footed we stepped out into the black, wet night. I shined a flashlight in the direction of the hanging moose. No bear there, everything looked fine. Then we heard the chattering noise that had alarmed Bob. I shined the light up in a tree on two grown martins out on a limb, having a heated family quarrel. Relieved, we returned to the tent and cleaned the majority of the mud off our feet before sliding back into our sleeping bags.

After three more long rainy days and nights in the leaking tent, the sun reappeared on the last day of moose season and we decided to head home. Breaking camp, we loaded the moose and gear into the boat. Bob untied us from the tree root and jumped in. I turned the key to start the engine. "Clunk," that's all, "Clunk." It had run fine six days earlier, with oil back in its system. I tried to pry the engine over but it wouldn't budge. My heart and my spirit sank to a new low.

Bob jumped out and retied us to the root. Without engine power our only salvation was to float downriver 115 miles to the closest village of Ruby for help. We cut some long poles to help guide the boat before pushing off again from the bank. As we floated slowly, we tried to avoid hitting the many sweepers (trees hanging over the river). I kept wondering what caused the engine to seize up. I knew water could do that, but the radiator was still full. I pulled the spark plugs and water ran out. Evidently rain had collected in the recessed center of the engine air filter around the bolt-hole. From there it had seeped in all week, funneled down into the carburetor, then into the combustion chamber through the open intake valves, causing a hydrostatic lock. Now, with the plugs out, the battery allowed the engine to spin over and pump out the water. Bob cleaned and dried the wet spark plugs and reinstalled them. We both prayed our troubles were over. I cranked and cranked before one cylinder fired, then another, spitting and sputtering. Finally the engine started. White clouds of smoke bellowed out of the exhaust pipe. When the white vapors cleared the engine smoothed out and we were on our way. I

129

told Bob, "That's about all the trouble I can handle; let's go home."

An hour down river we spotted Wiley Harris's boat and camp. Wiley is my rail-thin, camel-smoking neighbor who was hunting with our friend Smiling Sam. We stopped to warm up, have a hot cup of coffee, and swap experiences. Wiley had shot a young moose and planned to break camp and leave soon. He couldn't wait to tell us about Sam's encounter. "Without his rifle, Sam went about three hundred feet from camp and was hiding behind a clump of trees. He was squatted down with his pants down around his ankles when he heard a "WOOF." Looking up, he saw a young brown bear, sixty feet away and ambling toward him."

Then Smiling Sam jumped in to finish the story. "That bear scared the stuffing out of me. I tried to pull up my pants and run at the same time. I couldn't; I just fell down yelling, 'Wiley, get your gun. There's a bear chasing me.' My antics scared the bear away before Wiley came out. I had to change clothes and clean up."

Wiley added, "That bear also scared the smile from Sam's face for hours." We all had a good chuckle.

I told Wiley I figured I had enough gas to make it back to Tanana, and I planned to stop there for more; but to keep an eye out for me, in case we don't make it. Bob and I took our leave and continued the last few miles to the mouth of the Nowitna before turning up river.

Twenty minutes short of Tanana, the engine died in the middle of the Yukon River. Our boat was heavy with moose and the return trip was all upriver, causing us to burn more gas. Bob and I grabbed the oars and paddled for the closest shore before we were swept too far downriver. We reached the bank and tied the boat up to a rock and there we sat.

Three hours later we heard a boat motor approaching and we were prepared to flag it down. Around the corner comes Wiley and when he saw us, he motored over. He siphoned a few gallons of gas out of his fuel tank. I thanked him, since that would be enough to get us to the village. Wiley lit a Camel before waving his goodbye. As we approached Tanana, not seeing the barge stuck fast gave us hope. The gas station was open and I was able to buy gas and filled our tanks for the remaining trip home. We left late that afternoon and entered the mouth of the Tanana River for the last 225 river miles back to Fairbanks.

I wanted to reach Manly Hot Springs, seventy miles upstream, before nightfall. We were twenty miles from our destination when the sun slid down the backside of the far mountains. It was a cool clear night and the moonlight was bright enough to continue running the river. Minutes later a patchwork of scattered clouds drifted over and eclipsed the moon. Total darkness caught me running thirty miles an hour midriver and I couldn't see a thing. The next thing I felt was a jarring bump as my moose-heavy boat bounced on something and slid thirty feet to a stop. Killing the engine I located a flashlight and peered over the side of the boat into one inch of water. I stepped out of the boat and immediately sank past my ankles in mud. In an all-out effort to get back in the boat and free myself from the vise-like suction, my boot came off. There was no way we would be able to push or pry the mired down boat off the soft and sticky river silt of a submerged mud-bar somewhere in the middle of the three-hundred-yard wide dirt-laden Tanana River. Held prisoners in the confines of my heavy eighteen-foot boat, Bob and I were still a long 175 miles from home. To make this seemingly hopeless situation even worse, the water and the near freezing fall temperatures were dropping by the hour. The Tanana River would turn to ice in the coming weeks. All the intelligent river hunters had returned to the warmth, safety, and comforts of family and home, while we sat all alone and chilled, perched on that pile of mud. We were at the complete mercy of unpredictable and often-cruel Mother Nature. We were cold, wet, distraught, demoralized, and terribly miserable on this the bleakest of black nights. I fully realized that this could be terminal.

I prayed my heart out for hours, asking God for an inkling of what to do or for any kind of help. I promised God, myself, and Bob, "If we can just get out of this latest predicament, I'll sell my boat and give up hunting." This was the worst of all bad things that can happen on the river, other than drowning. All those negative thoughts I had earlier up the Nowitna River returned with a much deeper sense of abandonment.

We spent a miserable night in the crowded boat, shivering and discussing our plight. Not one solution came to mind. We needed outside help to survive this last ordeal. In the morning as the light of day began to creep over the mountains we both grew desperate for a cup of hot coffee. Since we were out of drinking water I scooped up a one-gallon can of silt-colored Tanana River water. After taking a look at the muddy contents, I was ready to throw it back. Bob stopped me, being even more desperate for coffee. "Let it set a while, till the floating particles settle to the bottom." An hour later I poured the murky water off the top into another pot to heat for coffee. I measured an eighth

132

⤳ *Exposing Alaska* ⤳

inch of mud that settled in the bottom of the first can. When the can of murky water on the Coleman stove came to a boil, Bob added a double portion of coffee grounds. I remember saying, "I can't believe we are going to drink this stuff." But we did. It tasted terrible, like coffee-flavored muddy soup, but it was hot and it helped . . . a little.

We spent the fretful day like two prisoners cramped in a tiny cell. That evening my ears perked up to the welcome sounds of a barge coming up river. When the captain saw us waving and signaling our distress, he stopped. Staying twenty-five yards away in the deeper water, workers lowered a rowboat. Two men rowed over, close enough to ascertain our situation. They were carrying a load of lumber to Manley Hot Springs and the man informed us this was the last barge run of the year. After negotiating a price they returned to their barge, picked up a long cable and returned. I attached the cable to the front end of my boat. They went back to the barge, put it in gear and slid us off the mound of gluey mud, and I released their cable and waved my gratitude. My prayers had been heard and answered.

I removed hands full of mud from my jet unit as we floated before starting the engine and again we were heading home. Several hours later, as the sun rolled down the backside of the far mountains yet again, we stopped and made camp. We did not want a recurrence of the previous nightmare. In the morning, neither of us wanted a cup of Tanana-mud coffee or another peanut butter sandwich, so we cranked up and went on home.

This ten-day hunt, which we had planned to be our best ever, turned out to be our worst. It cured me of jet boats and the desire for river running. No longer do I enjoy the

133

stress and the challenges of water highways or the frustrations of being stranded out in the wild. When thought and visions of succumbing to nature's raw power dance around in my head, I cringe. A return to the safety of established concrete roads and dirt trails brings peace to my heart and soul.

Days after returning home, I called the Daily News-Miner and placed an ad: Jet-Boat for Sale—Cheap. I knew that with winter fast approaching this would be the worst time of year to sell a boat in Alaska. However I did get one call from a heavily whiskered, grizzled sourdough named Ted. He admitted, "I search the ad section looking for the real bargains." He took a long, hard look at my jet boat and offered me a portion of my low asking price. After swallowing hard to suppress a counter offer, I kept my promise and let him steal my boat, and I wished him well. A tremendous relief flowed over me as that Valco dreamboat that caused me nightmares was towed out of my driveway under the season's first snowfall.

The city of Fairbanks was named in honor of Charles W. Fairbanks, vice president of the United States (1905–1909.) He never set foot in Alaska.

I Dove into the Chitina River

Mike Blackwelder

Everyone knows that the Chitina River is one of Alaska's most treacherous rivers. Folks say that the Chitina has never yielded up a body, meaning anyone who falls in the swift, silt-laden, frigid glacier-fed river is never seen again. But like many other Alaskans, I accept the challenge each year and go to Chitina to dip salmon for our winter fish supply and enough to smoke into tasty treats.

My tobacco-chewing neighbor Fred and I left early Friday morning and drove the 315 miles to Chitina and on to the campsite at O'Brien Creek. After crossing a narrow bridge we hiked several more miles along the deserted railroad track in Wood Canyon. Then, Fred and I slid down the rocky two-hundred-foot path to the turbulent and murky waters of the river. We sat perched on a rock, dipnets in hand, hanging out over a back eddy while a stiff wind blew dirt and sand in our faces. An hour and a half later, I counted twenty-one choice red salmon we had plucked from the river. This was my kind of fish getting.

I had brought an inflatable raft down with us in case we needed one. The next thing I knew a strong wind picked my raft up and floated it out in the river. I didn't want to lose the raft, so I took off running the way the current was flowing and dove in. The strong current took me right down. I had my boots on and they quickly filled with ice water. Being a strong swimmer I made it up and sucked in air. Then I spotted the raft and swam toward the twenty feet of rope trailing it. The river took me under and again I fought my way up. This river was just crazy. I knew I had to get to the raft or I would soon drown. I kept swimming with the current and being dragged down and fighting my way back up for air. Finally, I saw the rope and grabbed it. I went back under but kept pulling on the rope until I got to the raft and was able to get in. My body felt numb and I was spent. Someone in a riverboat saw me go in and rushed right over to save me but I was already in the raft so he towed me back to my rock.

Back on shore a neighboring dipnetter told me, "That was a dumb thing to do, diving in the Chitina. The day before a fellow slipped in the river and he never came out." I felt blessed to have survived this calamity and promised to never dive in after anything in this unforgiving river again.

The Facts from Prudhoe Bay
The eight-hundred-mile Trans-Alaska Pipeline is forty-eight inches in diameter and cost about eight billion dollars.

The pipeline crosses eight hundred rivers and streams and three mountain ranges. In early 1997, the pipeline was handling about 1.4 million barrels of oil (one barrel of oil equals forty-two gallons), condensate, and natural gas liquids per day. It takes about six days, traveling at just over five and one-half MPH for the oil to make the trip to Valdez.

Big Game Hunting in Alaska

Our Sheep Hunt

Bill Griffin

After weeks of planning, we grew more anxious as the day of the hunt arrived. We fueled up early and loaded all the gear into my plane. When I received the okay from the tower, we taxied down the runway, picking up speed. At just the right moment I pulled back on the throttle and we were airborne. Flying in clear skies, we headed for the Brooks Range.

My friend George grew quiet and for the duration of the trip he kept his face pressed against the window, just enjoying the scenery. On arriving at the peaks, I flew around the mountains, hoping to see sheep and determine where we would hunt. George spotted eight rams high up on one of the mountain and wasted no time in pointing them out. I circled back to the base of the mountain and selected a landing spot on a gravel bar. We set up our tent beside the plane and roasted hot dogs over our small campfire for supper, before turning in early. We knew we would need all of our energy just to get up to those sheep.

The next morning, after two cups of thick coffee and a large bowl of half-cooked oatmeal, our hunt began. We had an easy walk the first two hundred yards to the base of the mountain and then the tough climbing started. Up we went, fighting our way through scrub brush, crawling over rockslides, and clinging to rocky cliffs.

∽ *Exposing Alaska* ∽

Several hours later, we crested a layer of rock that had signs of a game trail embedded in it. We followed those tracks and came upon a grizzly bear feeding just off the trail about a hundred yards in front of us. So we sat and waited for that grizzly to get out of the way so we could continue. This was a real thrill for me, since I am a wildlife biologist. We sat there with our guns on that grizzly, not knowing whether it would charge us. Finally, after it had its fill, it walked away and disappeared. We continued climbing and when I peeked over the next ridge, there was a band of sheep. I shot and killed a really nice trophy, a full curl ram; it was just a lucky shot. George missed his ram and didn't get another chance because all the sheep vanished.

We butchered my ram and George put the meat in his pack and I put the sheep head and the hide in my pack. Now as we started down the mountain the sun slipped behind the far mountain. I said to George, "You remember there was a grizzly close by when we came up here. Now it's pitch dark and we both have fresh blood all over us and our backpacks are full of fresh scented sheep meat, so keep your eyes open."

This was a really dangerous situation and we both grew tense and concerned. A grizzly could smell us and attack before we could even see him. We were especially fearful when we passed the spot where that grizzly was earlier. We didn't have another choice so we just cautiously walked on through, all the way down to the river and to our camp.

If I did something like that today I'd be worried sick, but back then I was just half-worried.

138

～ Exposing Alaska ～

Dall sheep are white mammals with amber horns, yellow eyes, and black noses and feet. Males weigh about two hundred pounds and both sexes stand between three and three-and-a-half feet high at the shoulders. The horn clashing for which rams are so well known does not result from fights over ewes, but rather is a means of establishing a social hierarchy. Biologists estimate there are about forty to fifty thousand Dall sheep in Alaska.

I Wanted a Moose

Mark Neidhold

The year 1995 was going to be different, not a photo-copy of all the other unsuccessful years of moose hunting. Numerous scouting trips throughout the summer were instrumental in selecting our hunting area. We picked our spot after finding a variety of fresh moose prints pressed deep into the game trails. Several piles of still-steaming moose nuggets suggested that this hunt had the earmarks of success. Come September, John and I felt fully prepared and the adrenaline began pumping through my veins. My very own, prize-winning set of antlers hanging from my mantle captivated my imagination.

This was our plan: after the weekend hunters had left, my husky partner John and I would head up the river and hunt the last three, midweek days of the season. I picked a brawny partner to help with the heavy work because I'm a slight built man. The one thing we had not taken into con-sideration was John's wife, Dorothy, and the arrival of Jacob, their adopted two-week-old baby son. No way was Dorothy going to let John run off hunting and leave her alone with the added work and responsibility of their heaven-sent bundle of joy.

On such short notice I did not find a replacement with whom I was comfortable. I grew up in Alaska and was

∽ *Exposing Alaska* ∽

keenly aware of the risks of hunting and being in the wilds alone. However, my hunting instinct was acutely aroused and by God, I was going after a moose. I did file my exact location and expected time of return with my sister Katie.

It was cool the morning of the hunt when I arrived at the river. I slid my lightweight boat from the trailer into the clear water. I loaded my hunting gear, supplies and my canoe before cranking up and motoring up river to get my moose. The early stretches of the river I knew well, and I stayed clear of shallow water and sandbars. Light was fading as I reached my preselected main campsite. After unloading the night's necessities from the boat, I set up my tent, then prepared a simple dinner by flashlight: a can of beef stew heated on my one-burner Coleman stove. No campfires this week; I did not want the unfamiliar sight of flames or the sounds of crackling wood or the smell of a fire to frighten off my moose.

I knew from past experiences that the river was dropping in the late fall days. The return trip downstream would be more difficult, because the whole idea of this hunt was to return heavy with moose meat. So before bed, I placed a marker stick in the river to indicate the river level and made a mental note of a minimum level at which I must head back, moose or no moose. The night was beautiful and quiet, perfect for peaceful slumber. The only noises were the soothing sounds of a quick flowing river, an occasional bird chirp and the numerous slaps of a beaver's tail.

I awoke at 8:00 in the morning, late by hunting standards, but this day was still a day of preparation. First, I made improvements in camp by removing cooking and food storage items from the sleeping area for my safety if a hungry bear came by. Then, I covered everything with tarps to stay dry in case of rain.

∽ *Exposing Alaska* ∽

Next I unloaded my canoe from the boat and set out to prepare the area I would hunt. Paddling three quarters of a mile, I portaged over two beaver dams before reaching an oxbow lake. Then I paddled around the lake until I reached the hunting site that John and I had selected earlier. After tying the canoe to a willow tree I went to work with my father's machete, building a blind from green spruce bows and willows on the lake's edge. I wanted my concealment to look natural so as not to spook a moose. After finishing the blind I relaxed and the long wait began. I just needed to be patient and still so I would have a reasonable chance at calling a bull into my area.

As I waited quietly without distractions, I could hear my dad telling me, "Son, you are foolish for making the hunting trip alone."

Everything in place, I began my evening hunt by using an old moose shoulder blade and making a series of scrapes in the brush. Next at longer intervals I grunted like a bull. Then I gave my favorite call, the "coffee can cow moose call." I learned these techniques from my dad on previous hunting trips. As the evening light began to fade, I noticed the beavers working the lake becoming excited. Listening intently, I heard the answer to my call, a bull's antler scraping the brush somewhere out there. In previous years, I had never been this close. There were countless cows but this time it was a bull and he was responding to my call! After a time of breathless anticipation, I saw him: a tremendous bull. On top of his huge head, he carried that very set of antlers I had envisioned hanging from my mantle. This bull was skirting the lake, moving toward the blind I was nestled in. All I needed to do was be patient, wait, and pick the right time to shoot.

142

∾ *Exposing Alaska* ∾

Now big bulls are cautious. That's how they live long enough to grow large antlers. This bull would stop, check for anything unfamiliar, and then move ahead carefully and quietly. A seasoned hunter would have waited patiently for the animal to continue closing the gap but I was anxious and excited. I raised the rifle to my shoulder, but I couldn't control it, my adrenaline was running wild. The barrel seemed to move like a music conductor's baton, waving everywhere but on the target. Then I kept hearing this repetitive sound, like a whoosh, whoosh, whoosh and I realized it was the blood rushing through my ears. I could hear my father's words, "Mark, if you're not confident you can make a clean kill, you must not shoot."

I didn't shoot, and that trophy moose turned and headed back into the seclusion of the trees. I put the rifle down and frantically thrashed the shoulder blade through the brush inside my blind to encourage his re-appearance. My antics must have seemed comical to the bull moose. He did poke his head out again, just long enough to look my way and he apparently thought, not likely, because he slipped quietly back into the shadows. He was gone like a ghost. My adrenaline faded into a million thoughts of what I could have done differently and where I had screwed up.

Disheartened, and with daylight turning into blackness, I returned to camp for dinner and a good night's sleep. I was filled with mixed emotions. One minute I was frustrated with myself for messing up the best chance I've ever had to shoot. Next, I'd be all excited that I had actually called a bull into shooting range. Then I wondered what I could have done differently. Most of all though, I was proud of myself; without the confidence of a clean kill I didn't pull the trigger.

∽ *Exposing Alaska* ∽

The following days were not as exciting, but they weren't without promise. There were other bulls, perhaps the same moose that I had called earlier. I heard them get close, a muffled grunt or a carefully planned antler scrape, but none would show themselves. Each morning I was up early and stayed late at night. Only in mid-day did I take time to relax, returning to main camp for brunch and a nap. I gathered firewood, all the while looking forward to a nice fire after dark on my last night, and I spent anxious time preparing for the evening hunt.

The last day of the season came and I was frustrated and discouraged. I even started to formulate my explanation to my co-workers and friends for why I got skunked again. Then I heard it, at 7:30 PM with only four and a half hours before the season closed. Gentle footsteps trotted into the area. They stopped behind a stand of brush directly across the Oxbow Lake from me. Not hearing any antler scrapes, I assumed it was probably a cow. That would be okay though, because a bull might be following her. My eyes stared intently into the brush where the sounds stopped. Then, like in the hunting magazines, I saw his brow tines poking out from the edge of the brush. I saw nothing else, just the brow tines.

I reached for my rifle just as he stepped out. His head and shoulders were in full view. I froze in place, remembering all too well how my over-excited nervous system had ruined my chances previously. Again, daylight was fading and I knew my time was now, or I'd go home empty-handed again. My outstretched arms ached from holding the rifle still, in anticipation. I prayed the animal would move out of the brush into the open and commit himself.

144

Exposing Alaska

He didn't. I silently watched the moose slip back into the brush.

Thoroughly disappointed, I made one gentle cow call then a single tick on the brush with the shoulder blade. I picked up my rifle and pointed the barrel where I hoped he would come into the open. I must have been more convincing this time because he came charging right through the brush, ears back, nostrils flared, and hackles raised. This moose was definitely prepared for a scrap. He turned sideways and stopped. He didn't see or smell what he expected so he just stood there in picture perfect full view. This time I was mentally prepared and reasonably relaxed. I held my breath, carefully aimed, and pulled the trigger. Three well-placed shots from my Remington 30.06 and the moose dropped. I gave thanks for a bountiful harvest and felt proud.

It was 8 PM. The light was almost gone and the real work was about to begin. There would be no relaxed campfire back in camp tonight. I had chosen this particular area to hunt, knowing my own slight build and 140-pound frame would not be able to pack the animal far. I was three quarters of a mile from camp and my boat. The canoe would do the distance work but my moose was twenty yards inland. I still had to clean and cut the animal into manageable pieces. I had no actual experience, only stories and advice from my dad and friends. I looked at the massive animal and thought to myself, what have I done? And, I sure wish big John were here now.

By the dim glow of my Coleman lantern and a single six-volt battery-powered flashlight, I cleared the area of brush. The butchering process began by roping the topside legs to the nearby trees. I opened the hide and started to remove the front and back legs. The front leg wasn't too difficult, and I had it in a game bag with very little contamination from hair, leaves, and dirt. I felt I was getting the hang of butchering and started on the hind leg. The hip socket made it much more difficult than the front shoulder but I was managing. I was ready to bag the rear leg when I heard something in the brush that made my hair stand on end. Whatever it was stayed just outside the lantern light but I heard the brush breaking as it moved around me. So I shined the flashlight into the brush, searching for a clue to what was out there. A pair of big yellow eyes glared out of the darkness, revealing my worst fear, a bear.

My breathing stopped in mid-throat. I wanted to run, hide, and climb a tree all at once. Then reality kicked in. I knew I had to stay focused or that bear would have my moose and me for his supper. Shaking, I just stood there in the black of night. I couldn't even imagine a worse situation than being lathered in moose blood, working over a steaming carcass and having a bear circling me.

I had reloaded my rifle before starting work on the kill. Also, I had another full clip with me for a total of seven rifle shots and six more shots in the .357 magnum revolver on my shoulder. I shot into the air with the rifle and attempted to sound bold as I shouted into the night, "Get out of here!" The bear left temporarily.

After my heart quit pounding like a drum roll and my weak knees quit knocking, I was able to finish the hind leg. Next, I prepared to roll him over. I had heard the horror stories about trying to work around the gut pile so I put off opening the abdomen. With ropes tied into the surrounding trees and bushes I managed to roll him over. It took me almost thirty minutes. Each one of the moose legs outweighed my whole body.

Exposing Alaska

My mind, eyes, and ears were keenly tuned to any sound or sign of movement in the underbrush. The bear did return at roughly forty-five-minute intervals to check on the moose and me. Two more rifle shots into the air and much shouting discouraged it each time. Then I discovered I only had four shells left in the rifle and just the revolver to protect myself. The realization that I was exhausting my chances for survival forced another plan. I pulled the canoe up close to the kill and beat the aluminum framework with the plastic and aluminum paddles. The sound was quite impressive when combined with my shouting.

Working through the night and alternating between bear watching and cutting, I finished game-bagging my moose. That's when the heavy work began, loading my canoe and ferrying the meat back to the boat. Brunch the day before had been my last meal and all I had to drink was two diet sodas throughout the night. Exhausted and hungry, I paddled the meat back to the beaver dam in two separate loads. I had worked through the night and by 6:30 AM I was preparing to move the boat up to the second beaver dam. I managed to jump the first since it was only about eight inches high, and I docked against the low side of the second. By 8:00 AM, the meat and the canoe were loaded safely in my boat and I could finally relax and prepare some grub. My marker stick showed the river had dropped about six inches and was an inch below my mark, but I expected things would be all right.

The rising sun peeked over the mountaintop and there wasn't a cloud in the sky. I knew that the heat radiating from that ball of fire in the sky would put the meat at risk of spoiling, so I policed the area, broke camp, and headed for home. The return trip downriver was without incident, and

I arrived back at the landing where I loaded my boat on its trailer and headed home. I felt blessed with my harvest, my own bull moose, my winter's meat supply, and forty-three inch antlers to grace my mantle. I was elated that I didn't have to go home empty handed and explain why I got skunked again. And especially, I was thankful that bear didn't challenge my bluff and come charging out of the brush intent on a quick meal.

Alaska's first gold discovery was in 1848 by a Russian explorer.

My Best Moose Hunt

Tom Keele

Our freezer was empty and I grew desperate the last two weeks of the 1965 hunting season. Visions of a huge antlered moose carrying one thousand pounds of moose burger on the hoof kept me awake at night. I could picture a fat moose ambling out of the timbers and into the cross hairs on my 30.06.

Mike McBernie, my neighbor, had flown me over Indian Creek a week earlier in his two-seater plane, and we spotted several bulls. So I knew where they were; I just didn't know how to reach them. Walking wasn't an option. It was too far and traveling would be slow over muskeg and through swamps. A four-wheeler was a possibility but I didn't have one. Also vivid images of those four wheels buried to the hilt in a bog, miles from anywhere, kept appearing in my mind.

Then, Mike offered to loan us his John Deere crawler tractor and his four-by-six-foot sled. This was the perfect answer to how we could travel over the muskeg and reach this remote spot, plus haul our gear in and bring out our moose, if we shot one.

My wife Dorene, our son Bill, his buddy Tim, and I left our home in North Pole early the next morning and drove the two hundred miles to the Tok Cut-Off and down to the Chistochina River. I off-loaded the tractor and we packed the hunting supplies on the sled and began our cross-country journey. Heading northeast toward Indian Creek, we traveled all day. We successfully crossed swamps, bogs, and some nearly impassable terrain before reaching our campsite on the riverbank just as the day's sun slid behind the Alaska Range. By Coleman light, Dorene fixed a quick supper by heating two cans of beef stew and cutting into a loaf of her sourdough bread, while the boys and I put up the two tents and got out the night's necessities.

Early the next morning before first light, we were out hunting. Dorene led the way and I was right behind her as we walked around the first bend in the river. She stopped, mid stride, as her eyes focused on a nice bull, head down drinking in the river. She calmly raised her rifle, held her breath, and shot our first moose.

Bill and Tim had gone the other direction back to a meadow we passed through the day before. Within an hour we heard more shots. Both the boys ran back to camp shouting, "We got two moose down."

"That's great; Mom shot one too," I said. With the boy's help we spent the morning gutting and dressing Dorene's moose and loading the meat onto the sled. Then we headed to the meadow to tackle the boys' two moose when the tractor's engine overheated. The water pump seal gave out and coolant spurted from the pump. I shut the engine off and let it cool down. We trudged the mile over to the meadow and spent the rest of the day gutting and cutting up the boys' two moose, which we covered with a tarp.

The following morning I began removing the water pump. All I had for tools was a pair of vice grips and a crescent wrench. I had major trouble getting the rusted

⤳ *Exposing Alaska* ⤳

bolts to turn in cramped quarters, but I managed. For my efforts, I lost some blood from my scraped and skinned knuckles. It took about an hour to remove the water pump and there we sat.

Mike was going to fly over in a few days to check on us. We had some red surveyor tape so I tore off pieces and spelled out water pump on the sand bar so he would know what had happened.

The weather turned rainy and foul and didn't permit him to return for four days. The evening of the second day, I was out in the rain walking, checking the down river area when this big bull came out of the willows, not 150 feet away, heading for water. I shot him on the riverbank, and we spent the rest of the evening taking care of him. He turned out to be a real tough one, smelling strongly of the rutting season. His sixty-three-inch antler spread identified him as a senior moose.

The weather cleared on the morning of the fourth day and our friend flew over and read our sign. He dipped his wings in acknowledgement and left. Mike returned that evening to drop off a new water pump, which I installed with only one more bleeding knuckle.

I filled the radiator with Indian River water and started the tractor. After finding no leaks, we dragged the sled over and loaded up the boys' two moose. Then we went back and loaded my big moose on the sled. We returned to our campsite just after dark. We had four moose loaded up on the sled and we were seeing new bear tracks on the riverbanks and fresh droppings around. We took turns keeping bear watch all that night. It was spooky but the bear didn't show up. We never did see any bears; just the signs showing they'd been there, but that was okay. We left early the next morning.

151

Exposing Alaska

That was the best and most productive hunt I was ever on. I lost some blood and we had some stressful moments but all in all this was a memorable trip. I had my old bull ground up into hamburger and it helped feed us all winter. We shared many pounds of moose meat with neighbors who weren't so lucky.

How Big is Alaska?
- *Alaska covers 375,296,000 acres or 586,412 square miles.*
- *Its greatest distance east–west is 2,400 miles.*
- *Its greatest distance north–south is 1,420 miles.*
- *Alaska has 33,904 miles of shoreline: that's thirty-three percent of America's total shoreline.*
- *Seventeen of the highest mountain peaks in the U.S. are in Alaska.*
- *In 1880, Alaska's population was 33,426. By 1994, its population had grown to over 500,000 residents, making Alaska the third least-populated state. By 1998, Alaska's population had grown to over 615,900 residents.*
- *Alaska has a population density of slightly more than one person per square mile.*
- *Caribou outnumber people in Alaska almost two to one.*

Northern Lights—the Aurora Borealis has a curtain-like shape, and the altitude of its lower edge is sixty or seventy miles, about ten times higher than a jet aircraft flies. Fairbanks, Alaska is a good place for Aurora watching. The Aurora intensity varies from night to night, with the best viewing usually from late evening through the early morning hours, from September through April.

152

No Caribou Steaks

Paul Elbert

Steve Kerner, an eager man with a toothless grin, asked me to accompany him caribou hunting the following Saturday morning. It was to be a one-day trip because it was already October and winter was hard upon us.

Two hours before sunrise we left town, aiming to be at Eagle Summit, ninety miles north, at the first sign of light. The temperature hovered around zero and it began to snow, and of course, the mountain winds were howling. Steve had a new pickup but the heater was barely adequate to defrost the windshield, let alone warm us under such drastic conditions.

By the time we got to the summit, we were both chilled clear through. We drank the last of our lukewarm coffee and cruised up and down the road, hoping to intercept a heard of caribou crossing. About 10:00 AM I thought I spotted a small herd up a little canyon about a third of a mile from the road. I could only see about twenty dark objects moving among the trees. They looked big enough to be moose but there were too many of them, so Steve decided they had to be caribou.

At this point another pickup stopped behind us. Two young hunters packing brand-new 30.30 carbine rifles got out. They saw the herd we were looking at and asked if they could join in the hunt. I said, "Yes, if you follow instructions exactly and then all the meat will be split four ways." They agreed.

The stalk was simple. A low ridge not over one hundred feet high separated this little valley from the next. We went to the next valley and snowshoed up it until I thought I was opposite from where the caribou were. I stopped and told Steve to go on about two hundred yards and stop and for each of the two young fellows to go another two hundred yards.

When we were all spread out I gave the signal and we eased up to the top of the hill and peered over. The caribou were opposite Steve. A big bull came up to investigate Steve when his head appeared over the horizon. It was still very cold and blowing snow, just about as miserable a day as anyone had ever hunted in.

Steve raised his 30:06 to fire but found some moisture had frozen on his breach. He could not jack a shell into the chamber. He held his bare hand on it and finally thawed it out enough to jack a shell into the chamber and fire. Suddenly, the valley erupted as fifty or sixty caribou came boiling up out of the canyon. Steve emptied his gun and drove the herd right into the other two boys, who had worked back when they found they were beyond the caribou. They emptied their guns as the herd charged towards them. They turned the herd back towards Steve, who was reaching for shells to reload. Not finding any, he realized he had left his box of shells in the glove compartment of his pickup. The two young hunters reached for more shells and found that they also had left their extra shells in their pickup.

There stood fifty or sixty caribou (a boxcar full of meat) milling around between the three hunters. Finally the cari-

154

bou all pranced over the hill, heading for quieter surroundings. We returned to our trucks and then home without even so much as one steak to put in our frying pan.

It seemed incredible that three hunters could empty their guns into a large heard of caribou and not down a single one. But it happened. The weather, I admit was the toughest I ever hunted in.

I kidded Steve a few times after that by asking, "Got plenty of shells, Steve?" He must have been more sensitive than I guessed because he never again asked me to go hunting with him.

From My Cabin Door

Paul Elbert

My first bear was one to boast about. Even now, twenty years and numerous bears later, it still is the greatest of them all. I attribute this to one cause only, the phenomenon of beginner's luck.

A bear encounter was the farthest thing from my mind that spring day. What happened later is just one of the many surprising incidents that keep happening in a homesteader's life. I was transplanting the last few flats of cabbage plants into the field west of the house. It was 9:30 in the cool evening of May 20, 1945. Facing the golden glow of the setting sun sent shivers of exhilaration into my body and gladdened my heart. There is more to an Alaska sunset than any other sunset in the world. It lasts longer than those do in the Lower Forty-Eight, and its celestial beauty lessened my tiredness. I rested a moment after transplanting each flat and communed with nature's highlighter. Loneliness vanished.

I finished the last flat, rose to my feet and stretched. Suddenly a sense of uneasiness seized me, that sense that tells me when somebody or something is watching me. I spun around and glared at the unfriendly shaggy, brown hairy face of an enormous bear. Standing there erect and drooling, he was a good foot taller than a man. His paws sported two-inch-long claws and his front legs hung out in front of him like the burly arms of a hairy giant Sumo wrestler, raised and awaiting combat. He stood less than a hundred feet away. His eyes, narrowed in anticipation, seemed to be boring a hole through me. My legs began to tremble and I felt absolutely helpless. A terrifying scene flashed before me. I saw myself being dismembered, torn to shreds by the most savage of beasts, my limbs scattered about on the bloody ground. How painful would it be? How long would it take?

Physical combat was impossible. He was at least four times as heavy as I was and I'm a man of large physical proportions. I wondered if I might strike the hoe on its snout as it charged. Or maybe I could sink the sharp, six-inch hoe blade into its neck or back, so that it would eventually bleed to death. At least I would avenge my own death and prevent this man-killer from killing again. I also thought of drawing my knife. As the savage beast straddled me, chewing the life out of my face and head, I would thrust the knife up into his vitals and rip them open. That seemed the most logical procedure, so I held the knife in one hand and the flimsy hoe in the other.

The big bear must have read my mind. It also must have had a full stomach because it didn't attack. After a long, long minute it dropped down onto all fours. As he turned I could see the distinguishing hump on its shoulders, identifying it as a grizzly. It galloped out of the field into the woods.

I had gone from rapture in the beauty of an Alaska sunset to intense fear and terror. My heart beat like a drum and I perspired like a leaky faucet. I wiped the sweat off my face repeatedly, and eventually my heart slowed so I could again hear myself think. My legs felt like Jell-O; how-

Exposing Alaska

ever, I had enough curiosity to backtrack the bear. I found the path where he had entered the field and just beyond his trail I smelled the decaying flesh before noticing a large dead dog. Apparently the bear had eaten its fill before stalking me.

I knew bears were creatures of habit and reasoned that the grizzly would return the next evening. The dead dog lay near the edge of the field, just two hundred yards west of my one-room cabin. My only door faced that direction.

The next day, as evening approached, I made sure that my little 25.35 deer rifle held its full quota of seven shells. I moved my chair to the open door at 7:00 PM and waited. I passed the time by reading the short articles in Reader's Digest. As I turned a page or finished an article, I would look up. At 8:30 the big brown phantom glided out of the green spruce trees over to the carrion. I watched for a minute as the bear circled the dog, tearing off bites.

Suddenly, I felt panic again and deep concern. My gun seemed too small and the bear too big and too far away. I considered stalking him across the field to get a closer shot. But common sense rendered that impractical since it would put me in the open and consequently at a great disadvantage if the grizzly charged. I really doubted that my small caliber gun would kill that large grizzly.

Steadying my rifle against the doorjamb, I squeezed the trigger and hoped for a lucky hit. The bear went down, but got right back up. I fired the remaining shots rapidly and it seemed that each time I fired, the big brute buckled and fell. I had visions of seven slugs in his vitals. I reached for my box of shells. Only three remained. I inserted them into the empty magazine and jacked one into the barrel. By now the big brown brute had managed to flee into the spruce

158

Exposing Alaska

trees. I could no longer see it from my cabin door. I walked straight to the spot where I last saw it and there it lay under a spruce tree, gravely wounded. Its back was towards me but it kept turning its head, trying to see me. I took aim on its neck and fired. It sprang to its feet and galloped off and I sent my last two bullets after the fleeing target. The bear only lasted for a two hundred-yard dash. I heard its body hit the ground and the gurgle of blood in its throat. But it was still alive and I was out of shells.

I hurried back to the cabin, got my .22 rifle and went back to polish off the big grizzly with a shot between the eyes. On returning, I found the bear near death. It had fallen in the center of a small clearing in the woods. I worked my way around until I was at a right angle to his skull. I fired, but the great beast raised its head no more hurt than if a mosquito had bitten it. I fired again and again with the same negative results. But finally a .22 slug made it drop its massive head and lie still. As I watched, there were scarcely any involuntary tremors or spasms. I waited a respectable length of time before going in to cut its throat. It did not bleed, and then I saw that one of my bullets had severed its jugular vein.

I skinned my trophy and found I had a man-sized job on my hands. The big paws gave me the most trouble because I wanted the long claws left intact on the skinned-out hide. I later stretched it and tacked it up on my shed wall to dry.

The meat looked good so I cut off a couple of thick steaks for dinner. They tasted as good as beef. Because I lacked electricity and adequate refrigeration, I stored the rest of the bear meat in my covered, permafrost-cooled container, a three-foot-deep hole in the frozen ground. I

159

Exposing Alaska

also gave bear meat away freely to friends. I gave some to Mr. and Mrs. Nick Eidem, who were doing some graduate work at the University of Alaska Fairbanks. They liked it so much that they asked for more to share at their All Alaskan Game Banquet. I gave them a hundred pounds for their dinner. Later, several of the guests took the trouble to look me up to thank me and tell me how much they enjoyed the bear meat.

Shortly after that bear scare, I went into town and purchased a new high-powered Remington 30.06 and two boxes of 180-grain bullets.

That first grizzly bear was the biggest and the most impressive bear of my hunting career. It was the first game animal I shot from my doorway, although shooting from the luxury of my cabin became a habit. I can truthfully say that at least half of my game has been shot from my cabin including several moose and two wolves during the four years I raised chickens. My gun rack was nailed up close to the door and my guns were always loaded for bear, moose, or wolves.

Hunting Bears

Paul Elbert

Major Cathy was in charge of buying local produce for the Army's field issue and the commissary. That's how I got to know him. He had been a Florida farm boy and understood our problems. He gave us farmers a lot of consideration.

Because of our hunting conversations and my bear stories, the major made up his mind that he wanted a bear rug for his wife. He went into town and bought himself a new 35-caliber Marlin rifle. Then he asked me to help him get a good-sized bear. I told him to just come out to the farm the next evening after supper.

I had seen a black bear skirt the low field, slinking through the sparse bushes for the last two evenings. When the major arrived at about 6:30 I told him to back his car into the garage and just sit in it and watch for a bear to cross the lower field. He looked at me as though I was crazy. It took some tall talking to convince him that it was the safest way to hunt bear.

I didn't expect the bear to show till about eight so I went back to feed and water my chickens and get in the night's wood and do chores in general. I kept a sharp eye out for happenings along the lower field and, by chance, I just happened to be looking when the first flash of dull black appeared in the brush to the far right. I hurried down to the garage and pointed it out to the major. He was incredulous. He couldn't believe it was really a bear ambling through the bushes. He grabbed his binoculars and studied the bushes. Finally he put them down and said in blank amazement, "It is a bear!"

By that time the bear had crossed through the short brush and entered taller thicker bushes and was out of sight. I said, "Major, you should have shot." He looked mighty crestfallen and made some derogatory remarks about missing his chance to shoot his bear. I quickly assured him that the bear was simply looking for something to eat and would probably return in about an hour.

The long shooting distance, being roughly two hundred yards made the major want to get closer. Now on the lower edge of the field grew a ten-inch, bushy spruce tree. Previously I had taken a handsaw down and cut off enough limbs to leave an easy-to-climb ladder. I told the major to go down and climb that tree, which he did.

He waited nearly an hour up in that tree and later told me that the limb he was standing on had almost cut into his feet. His arms ached from clutching the trunk to keep from falling. He was very uncomfortable to say the least, but he waited in silence.

When he saw the first glimpse of the bear returning from its stroll, his pain vanished. The bear stood up 100 feet from the tree and the new .35 Marlin barked. The major had his bear (or so he thought). While the major climbed down the tree, the bear got up and ran away.

I was looking when the major shot. I saw the bear fall. I picked up my handsaw and whetstone and rushed down to help the major skin out the rug. When I got there the

Exposing Alaska

major was examining the spot where the bear fell. It had rolled about considerably, and the grass and moss were crushed as if a moose had bedded down, but there simply wasn't any blood. We couldn't find a drop of blood there or anywhere else. We circled far and followed each trail and no blood but no dead bear.

The major returned the next morning and the next evening but the bear did not show up. However, two nights later the bear did appear and I rushed down to the lone spruce tree, climbed to my perch for a better view, and when the bear was in the cross hairs, my new 30.06 spoke, and the slug broke the bear's back. It didn't move five feet. I bled it and then drove two miles to the nearest telephone. I called the major to come on out and skin his bear rug. I told him I couldn't be sure but I thought it was the very bear he had wounded. He wanted to know where he had hit it. I told him to come on out and see.

In about twenty-five minutes he skidded to a stop in the yard. He hurried down across that field to see that bear. When I showed him where his bullet had creased the bear's scalp and knocked him down, we both understood why there had been no blood. That convinced us it was HIS bear.

The major skinned the bear rug carefully. He sent it outside to a tanner and about six months later got back a beautiful bear rug. The major felt a little chagrined at not having actually shot the bear and also at getting it so close to town in a homesteader's back yard. He said he was going to make up a big story of his wild bear hunt while traversing cruel terrain in far-away Alaska for the folk's back home.

I told him I thought the simple truth would be more interesting to his folks back in Florida; just tell them that most any day a homesteader might shoot a bear from his cabin door.

163

Youngest Hunter Kills Oldest Buffalo

Douglas Elbert

This five-word headline in bold print, above my story in the *Jessen's Weekly* newspaper in Fairbanks Alaska, made me ecstatic. I nearly burst my buttons swelling with pride that October day back in 1950.

Alaska buffalo and the Big Delta farmers were destined to collide. Nine years before Alaska became the forty-ninth state, the Alaska Game Commission considered ways to avoid this inevitable conflict. The problem resulted from the buffalo doing what they were born to do—roam. They weren't content to stay on their ninety thousand acres of state land, the Delta Bison Range. The buffaloes were ravaging and trampling the crops, and the local farmers were getting fed up with it. The farmers' efforts to keep the buffalo from eating and ruining their produce included fences, chasing, yelling, fireworks and yes, blasting the buffalo with bird shot from both barrels of a twelve-gauge shot gun. All had proven fruitless.

Then, in the spring of 1950, the Alaska Game Commission came up with a possible solution: thinning the expanding buffalo herd in Big Delta. The commission initiated the first of what has since become a yearly buffalo hunt by issuing a special permit for a drawing to be held in August, along with the hunting licenses they sold. Earlier that summer, my dad, a farmer and an avid hunter and sportsman, bought licenses for himself, my two older brothers, and me.

Farm work was tedious and Dad worked us long hours in order to take advantage of the twenty-four hours of sunlight during our short growing season. Our hours and days were spent planting new seedlings, cultivating, and weeding. The weeks and months seemed to roll by until that special day in mid August. The five acres of cabbage plants had matured into giant heads. That day, one week after my sixteenth birthday, became one of the most memorable days in my life. Dad, my two brothers and I were busy cutting and crating two tons of prime cabbage for an Army order when the noontime whistle blew. We dropped our knives and hustled the 150 yards back to the house. We took turns dipping rainwater out of a wooden barrel and washing up in a chipped white enamel dishpan.

Seated at the table, Dad reached for the newspaper while Mom placed a large platter of fried chicken on the table. This was the newspaper we had anticipated all summer long, the one that listed the lucky twenty-five buffalo permit winners. Dad's solemn eyes scanned the list until suddenly his face broke into a grin, which continued spreading like wildfire. I could sense his excitement, but without a word he handed the paper to me, pointing to my name. When I saw Douglas Elbert, me, on the winning list. When it sank in that I was drawn as one of the twenty-five hunters with a permit to take a buffalo at Big Delta, I could hardly contain myself. I jumped two feet straight up in the air. In the excitement, my thoughts and dreams focused on becoming a great buffalo hunter.

Later that night, after the day's work and chores were behind us, Dad took me down to the lower field where my

Exposing Alaska

first-ever target practice preparation began. He had drawn a two-inch bull's-eye with a black crayon on a white paper plate and tacked it to a tree stump. My beginning efforts were spent trying to keep the barrel of Dad's 30.06 from waving around like a music conductor's baton. But after going through several boxes of shells in the evening practice sessions for a month, I progressed from missing the whole target to hitting it sometimes. I kept shooting even though my shoulder was taking a beating. Dad's 30.06 had a kick. Slowly, I got pretty good at hitting the target most of the time and the bull's-eye half of the time. Dad and I agree I was ready.

My scheduled big hunting day arrived on the ninth of October. The weather outside was cool, but I felt a warm glow inside as Dad and I left Fairbanks after lunch. We drove the hundred miles south and checked in at the Alaska Game Commission camp in Big Delta where I happily displayed my prized permit.

The envious game warden smiled and said, "You lucky guys. Okay, the way we work this is to send a plane up first to spot a buffalo and I will send you both out to get him." A short time later we hear the crackling radio report coming back from the plane. "There's a big bull in a field, down by the river, above Rika's Roadhouse." My blood was pumping as another warden guided us to the exact spot described by the pilot earlier. We drove to a large, lush, half-eaten, well-trampled field of peas and oats. At the far end was an enormous buffalo. He looked almost round with his bulging belly. He was just standing there, chewing his cud. Dad and I slowly crept to within fifty yards and, at 4:00 PM. I drew a bead for a perfect heart shot, fired, and— it wasn't so perfect! That elephant-size buffalo didn't even flinch. He slowly began to amble toward the woods that

166

∾ Exposing Alaska ∾

surrounded the field. I tried to quiet my pulsating nerves and calm my vibrating rifle while firing three more shots. He fell like a rock. I was ecstatic. I think I jumped three feet up in the air that time.

We later learned that the local natives had named this buffalo "Old Bill," because he was the oldest and largest buffalo in the State of Alaska. Notches in his ears identified him as one of the original bulls transplanted there twenty-two years before. His broken horn identified him as the Old Bill that had wrecked five jeeps for the army boys who tried to chase him off the Fort Greely's runway, by honking at him. "Old Bill" had also tangled with a big Buick and left four inches of his left horn imbedded in the twisted body. The Buick came out second best.

We also learned that earlier in the summer Old Bill had found this field of peas and oats that a farmer had planted; Old Bill had moved in and proceeded to chase any other buffalo away that tried to enter his territory. When the farmer came with a big stick to chase him away, Old Bill lowered his head and looked the farmer straight in the eye. This was the showdown; they stood staring at each other for several long moments. Then the farmer took a step back and surrendered all his rights to the crop of peas and oats. Old Bill proceeded to try to eat everything that grew on that five-acre field all summer long. He had grown so fat he could hardly walk.

We had no way to weigh him but the game warden and Dad figured by his age and size that he had to be over 2,500 pounds, probably the heaviest animal that had walked in Alaska in a long, long time.

The sun dropped from the sky at 5:00 PM and the smidgen of warmth vanished as the coolness of Alaska's October

167

Exposing Alaska

nights returned. Just then, two gaunt, bewhiskered fellows in tattered prospector's coveralls showed up and asked, "Ya'll want us to skin your buffalo and quarter him for yah?" Dad wasted no time in saying, "Yes," and he told them to take a good number of thick steaks for themselves.

Dad and I took a break and went to Big Delta to celebrate with a big juicy hamburger, and he told me how proud of me he was. He said, "Son, you just accomplished something that I and most other Alaskans have wanted to do but we never got the chance to shoot a buffalo. That was some good shooting, Son. He fell clean like you hit him in the heart."

The temperature was dropping outside the Roadhouse, but I felt very warm and accepted inside, especially since Dad had never showed emotions or expressed his feelings toward me before. I couldn't imagine how anyone could stand being any happier than I felt right then.

After our burgers we went back to the field and loaded Dad's pick-up with the buffalo meat, the hide, and the head. We thanked our two helpers and noted the large portions they took for their efforts. Dad said that was okay though, because they did the hard work and it sure looked like they needed to put some meat on their bones and fill out their sunken tummies. We waved goodbye and departed for Fairbanks.

Fifty long years have passed since my big day, but I can still remember it all as if it happened yesterday. It's still my most favorite and memorable Alaska experience. I don't recall what happened to the buffalo hide or the trophy head, but I can still recall how good that grain-fed buffalo tasted. I didn't pursue becoming a great hunter and I've never killed a moose, bear or even a rabbit. However, I vividly recall the thrill of shooting Old Bill, the biggest and the heaviest critter that had walked in Alaska in many a year.

In 1928, twenty-three bison were shipped from the National Bison Range in Montana to Delta Junction and released. In 1996, there were roughly one thousand bison in Alaska. The largest land animal in North America, bison can weigh more than a ton and stand six feet at the shoulder. Even though they may appear slow and clumsy, they have great endurance and can run as fast as thirty-two MPH. Alaskan bison have been known to live up to twenty years. In the winter, bison require windblown areas that keep forage free of snow.

Bison cows usually give birth to a single calf in the spring. Bison calves are able to stand when only thirty minutes old. Within three hours of birth, they can run and kick their hind legs in the air. After about six days, calves start grazing.

Triumphant in the Trophy Pasture

Don Elbert

My sixty-five-year-old father, Paul Elbert, was a man of large proportions with a young heart. He had grown sweet on an adventure-spirited fifty-one-year-old photographer-writer named Frances Vickery. One fall evening in 1964, he went calling on his lady friend. He brought two thick moose steaks from his recent hunt with Sam Gamble, instead of his usual farm-fresh vegetable offerings. Frances' blue eyes sparkled with pleasure as she invited him to stay for dinner. While the steaks sizzled, Dad relayed the details of guiding Sam in shooting this moose. His stories prompted a request to take her moose hunting. "Please, Honey, I'm writing a book about the highlights in my life, and I know a hunting trip with you will be a most interesting chapter."

Dad made the arrangements and on the following Sunday, September 15, they climbed aboard an Alaska Air Services Cessna 180. Forty-five minutes later the pilot landed on a rocky strip near the Savage River, just north of Mt. McKinley Park.

The campsite Dad chose was on a shelf that overlooked the four-hundred-yard wide, three-mile-long valley below. The place Dad had named the Trophy Pasture. He spread their sleeping bags several feet apart on dry moss. The weather was pleasant, and the forecast spoke of clear skies ahead. However, Dad brought an eight-by-eighteen-foot strip of thin plastic Visqueen along for a covering just in case the weatherman spoke with a forked tongue.

They enjoyed a simple evening meal of hot dogs, corn chips, and an apple. After the night's chores were finished they sat on a log in front of a crackling campfire and enjoyed the evening performance of the northern lights. Frances confided her feelings. "I'm excited and love being out here with you, Honey, watching the Aurora dance across the sky and listening to the quietness of nature."

Another contented hour zipped by and the two slipped into their sleeping bags. Although he should have known better, my father proceeded to lose all her admiration when he explained why he separated the sleeping bags. "So that one of us could hopefully protect the other in case of a sudden midnight bear attack." This frightening image did absolutely nothing to make her feel safe under his protection. In spite of bear assault thoughts, they continued viewing the overhead entertainment and finally drifted off to sleep. An hour later, Dad was startled awake by cold raindrops racing down his face. He quickly reached out and pulled the plastic visqueen over them.

They arose with dampened spirits at 5:30 that drizzly, chilly Monday morning. Frances grumbled constantly as they climbed into rain gear. They hiked on down an old cat trail in the pre-dawn light. Fifteen minutes later Dad spotted a white object on the hillside and whispered to Frances, "Hon, there's something up ahead." As they neared the spot, they noticed movement. Dad led the way as they crept within one hundred yards of the mysterious object. Stopping in their tracks, they watched a big straw colored Toklat grizzly bear digging up the tundra, attempting to cover up a moose kill. An excited Frances exclaimed, "Paul, shoot!"

The rain had coated the big end of the Bosch and Lomb scope on Dad's Winchester model 70 rifle. He quickly wiped the scope, sighted, and squeezed the trigger. At that same moment the grizzly reached for another claw full of tundra, which lowered his body, causing the 220-grain slug to strike high on his shoulders. The beast bellowed, bit at his wound, and fell over on his side. He got right back up and limped into a birch thicket.

A now fully energized Frances pleaded, "Let's go after the bear. You wounded him, Paul, and he can't go too far." Dad flatly told her, "Oh, no. No sane person follows a wounded grizzly bear into a birch thicket!" Francis accepted his wisdom and they returned to camp for breakfast and rest. She fried Spam and buckwheat cakes over the rusty Coleman stove, while fresh coffee perked.

The cold rain continued, at times turning to snow. Dad's thoughts returned to the wounded grizzly. After a rest, he went back and tramped the hillside, but he hunted in vain. The terrific stamina of this Toklat grizzly had borne him to a safe hiding place, further away.

Dad lumbered back to camp as the storm intensified, and a glance in Frances' direction revealed a scowl of displeasure. Camping under an eight-by-eighteen-foot piece of loose Visqueen did little to provide adequate protection. The foul weather had made it entirely too miserable for any pretence at enjoying their outing.

Dad knew from previous years of an abandoned trapper's cabin half a mile away, across the Teklanika River. He was successful in convincing Frances that the cabin was the smartest move they could make to reach a shelter. They held hands as they briskly sloshed the six hundred yards to the river's edge with the rainwater pouring off them. Here, Dad being the gentleman, bent down and invited Frances to climb on his back, and she did. Dad waded in the murky water, carrying her piggyback across the forty-five foot wide river. This was a slow and strenuous task because the water was three feet deep and the current faster than normal. His load combined with the slippery stone footing made the crossing treacherous. He slipped as he attempted to gingerly step out of the water on the far shore. Frances fell backward on a sharp stone and screamed in pain. Soaking wet and crying, she blurted out several unprintable remarks. Dad swallowed his immediate response, knowing it would just bury his cause. He apologized in an effort to console and regain her confidence. Dad supported her as she hobbled the last two hundred yards to their shelter.

On arrival they discovered the cabin nearly concealed under a matted brush covering. The door had fallen off the hinges. It was a small, low ceilinged, rotting log structure. The latest occupants appeared to be porcupines and squirrels and they had literally torn up the place. The remains of a bag of flour covered the floor like a coating of snow. The inside felt clammy. It was cold, dark, and reeked with the musty smell of decayed slop bucket garbage. There was a rusty but serviceable Yukon stove, some dry wood, and a little oil that helped in starting the fire. As the wood crackled loudly in the stove the cabin did heat up and by bedtime Frances was beginning to dry out. This rise in temperature did wonders toward thawing her frigid disposition. She wrapped her arms around Dad's neck and

Exposing Alaska

apologized for her hard words while giving him several long, wet kisses and a big squeeze.

Tuesday morning, everything outside the cabin was painted white with a two-inch dusting of snow, and the storm continued. It appeared the hunt was doomed to complete failure. However, by mid-morning the clouds separated and the sun rays reached down and touched earth. An hour later, the snow was mostly gone and a recuperated Frances came out to view the beauty of Alaska's wilderness. She enjoyed the simple things of nature and took pictures of the thin black spruce trees that stood like silent statues guarding the cabin.

At noon, Dad carried the tentative Frances back across the Teklanika River without a mishap this time. She said, "I wish there had been a third party to photograph my piggy-back-ride on the back of my sixty-five-year-old white-haired guide." Then she hugged him.

Dad rested there while she spent the next hour viewing and photographing the freshly washed faces of the rocks. Never before had Frances seen so many twisted and contorted stones in such a variety of colors and shapes.

By one o'clock, they had returned to the first campsite overlooking the Trophy Pasture. The valley was mostly covered with two to six feet tall, buck brush. Dad studied the terrain below with binoculars while Francis, now in a better mood, busied herself making lunch.

Suddenly, Dad's solemn face lit up like a hundred-watt light bulb; he turned and whispered in a heavy voice to Frances, "antlers coming!" She looked in the direction Paul's large hand pointed and gasped when she saw them glistening in the sun. Spread out like huge sails, the wide white antlers appeared to float above the brush.

174

Then the bull moose became apprehensive and veered off at a brisk walk. A noise, scent, or glimpse of them had spooked him. Dad gave his irresistible cow call, but this bull paid him no heed. Dad urged Frances to shoot before the rapidly departing moose was out of sight. Dad stood behind her coaching, "The bull is four hundred yards away now. Your bullet will arch, so hold the cross hairs of your scope twelve inches above and twelve inches ahead of the front shoulder. Squeeze the trigger gently as soon as you get the crosshairs on target."

A moment later a shot rang out. The big bull took two more steps and collapsed in the dense brush. Frances looked at Dad in astonishment and then her face lit up as she exclaimed for the world to hear, "I did it, Paul! I shot a moose!" Dad beamed with pride, because even though the weather had been nasty, he had guided his lady love on a successful hunt. He felt a portion of her faith in him returning.

The moose fell halfway between a lone cottonwood tree and two of the larger spruce that decorated the tundra. Dad marked the spot in his mind as he turned away to retrieve his pack board, which contained a saw, rope, a honing stone, and skinning and butchering knives. He also grabbed his camera because pictures are the sought-after proof of his lady hunter's prowess.

They hurried to the spot Paul had etched in his mind. They scoured the area, but no moose could be found. Dad climbed a cottonwood tree to a height of fifteen feet: no moose. He then went to a nearby spruce tree and climbed equally as high: still no moose. He and Frances made wider and wider circles without any sign of a moose. They peered deep into the brush, behind scrub black spruce trees and in every large depression in the muskeg. They traversed the valley well into the late afternoon, but they looked in vain. By now, both Dad and Frances were sweaty, hungry, worn-out, and bone-tired.

My disheartened dad said, "I think the moose must have recuperated and trotted off when I turned away to get my pack board."

All that previous joy drained from Frances and her face drooped as they called off the search. Two mighty discouraged hunters dragged themselves back to camp that evening. They replenished the campfire and sat quietly drinking a cup of whiskey-enriched coffee.

Squadrons of geese and cranes honked and squawked as they circled above and the fog raised from the nearby hills. Then, over the hills droned the whine of an airplane. Soon a Cessna landed and two grinning hunters disembarked. They were Al Peterson Sr. and his twenty-four-year-old son Al Peterson, Jr., who had recently completed his tour of duty in the Marines.

The next few minutes were spent in small talk, mostly trying to get acquainted and discover each other's hunting plans. Then Dad told of Frances' bull and that he figured it had gotten up and run off. He asked the Petersons if they would take to the air again and fly around the cottonwood and the spruce trees to see if the bull was still in the vicinity.

The Petersons made two passes over the area before slipping back in and landing. Al Sr. had spotted the moose on the first pass and Al Jr. had pinpointed it on the second pass. It lay near where Dad said it fell. Dad and Frances were so close; they had walked within thirty feet of it several times. But the five-foot brush was very thick in that area and obscured the moose completely.

The Petersons now had an interest in our moose and offered to dress it and carry it out to the landing strip for half. To this Dad and Frances readily agreed and the four of them scampered off to begin work on the moose. Frances, her face just beaming, insisted that pictures of her with her moose had to be taken first. Then they watched and rested as father and son dressed and quartered the moose. Al Jr., a fine specimen of strength and endurance, quickly packed that moose out, one quarter at a time.

After a lengthy rest back in camp, more conversation, and several cups of strong coffee the Petersons departed, taking to the air with their half of the moose. Hunter's instinct made them circle the area in a one and then a two-mile radius. Then they dropped back in for a third landing. After parking the plane, they brought out guns and gear for a hunt.

The younger Al rushed over to Dad and Frances and exclaimed, "I spotted an albino bull moose in the next valley. Do you want to join in our hunt? We got half of your moose. I think you should get half of ours, if we are successful."

In agreement, the four started out together, but Dad and Frances could not keep up with the youngsters. So they decided to fork off and cover the intervening valley, in case the white bull evaded the Petersons.

By mid-afternoon Dad determined that there was no further need to cover the intervening valley. They had not heard the shots they would have heard if the Petersons had found the white bull just over the hill. They turned back towards the landing strip and were just coming down off the hill when Dad boomed, "antlers ahead!" A mile away, a bull was browsing downwind in the brush. Dad pinched his nose and gave his favorite cow moose call. Immediately the big white antlers rose high in response, not to be lowered again as the bull trotted straight to the call. His path was parallel to the landing strip and right down the middle of the Trophy Pasture. When the bull was opposite the end of the runway and still two hundred yards away, Dad raised up from his crouch. The moose stopped in his tracks to stare at this strange sight. Dad placed the crosshairs of his scope on the bull's forehead and pulled the trigger. The 220-grain bullet sped to its mark and the moose buckled and collapsed, stone dead in an open meadow of muskeg. Frances watched from behind, making mental notes of every detail for her forthcoming book.

They hurried to the moose and took a whole roll of pictures and then Dad properly bled and gutted the moose. The sun was setting when they returned to camp for a needed rest just as the Petersons were arriving. Al Jr. loudly proclaimed his successful kill of the albino moose.

Excitement again raced across Frances' face. She had an expensive camera and could take exclusive pictures of the albino's demise. Albino moose are a rare freak of nature and most albinos observed have been cows.*

Wednesday morning, Dad and Frances started out early to be at the scene of the albino kill before the skinning started. They didn't make it. The Petersons had one side skinned out when the camera lady arrived. Because this was such a trophy they planned to skin out head and feet

---

* It is common knowledge that the big albino moose with modest antlers (that was exhibited at the University of Alaska Museum until 1983) was a large cow moose. A taxidermist had mounted the antlers taken from a smaller bull moose.

∽ Exposing Alaska ∽

and have a full mount, even though Dad advised that the taxidermy bill would be over a thousand dollars.

Frances used up all her film, two rolls of colored and three rolls of black and white. The albino bull was a three-year-old with a forty-inch antler spread. The white hide was dappled with several small brown spots and a large brown spot adorning the left flank. Its antlers were the smallest of all the sixteen kills Dad had seen taken from the Trophy Pasture.

After photographing the albino, the two returned to Dad's kill late that evening. Frances held a three-cell flashlight and Dad skinned carefully until he removed a perfect hide. For years my father had hoped to shoot a big bull through the head so he'd be able to skin the hide off, hole free. He envisioned having it tanned and a winter cabin scene etched onto it.

Thursday was spent getting Dad's moose meat to the landing strip. Frances helped carry and tug as best she could, but my aging, muscular Dad carried the bulk of it and he was ready to drop. Their steps were slow as they returned to their primitive camp hungry and thirsty. Frances asked, "Hon, will you go to the Savage River for water; I'll make you some fresh coffee." Of course my fatigued father replied, "Yes, Dear," and began walking heavily toward the river, a thousand feet away. Eyes half-closed, shuffling, stumbling and carrying nothing but a water bucket, Dad dragged himself along. As he neared the river, he noticed a light colored animal slipping through the brush ahead of him. Then, Dad's heart skipped several beats and his legs began to tremble as a three-year-old grizzly bear emerged from the brush and stood up a hundred feet away. Their eyes locked as they stared at each other.

179

Exposing Alaska

Too weak to run now, no gun, not even a knife to defend him, he faced the beast. The fear that seized him luckily was not the paralyzing kind. Reaching deep for the courage, Dad walked menacingly towards the grizzly, waving the water bucket and hollering in his most persuasive manner, "get the heck out of here, you rascal!" The grizzly stood his ground until Dad had advanced some forty feet; then he turned his head, dropped down, and walked briskly back into the brush. Dad inhaled a deep breath and emitted from his heart, "Thank you, Lord!"

The next day, Friday at noon, the Cessna 180 returned to transport the two successful hunters and their moose back to Fairbanks. Even though the pilot insisted on leaving several hundred pounds of meat until the next day, the plane was still overloaded. He barely cleared the treetops on takeoff.

In recalling the hunt a few days later, Dad said, "That trip with Frances was plagued with foul weather and some bad times but overall it ended in triumph. Frances beams whenever I join her for a feast of moose steaks. And her smile is the brightest when showing off her prize moose photos of three successful hunters in the Trophy Pasture."

(Paul Elbert and Frances Vickery were married two months later, but that is another chapter for Frances' book.)

Brown and grizzly bears are considered to be the same awesome creatures. Brown bears live on the coast and grizzly bears live in the Interior. Kodiak brown bears are the largest because they eat so well during the salmon run. Bears are curious, intelligent, and potentially dangerous animals.

Hunting Wolves

Paul Elbert

The first time I set out to find a wolf den I invited a neighbor boy, fifteen-year-old Bob Taylor, along. Bob owned a press camera and had gained his expertise while being the photographer for his high school. He readily accepted and was eager to take pictures of a live wolf in the wilds of Alaska.

We departed Fairbanks early on April 18, 1966, in my old three-quarter-ton Ford pickup, towing a trailer carrying my Farmall tractor. The year before I had converted my farm tractor to a track vehicle for traversing the wet spongy tundra, hopefully without getting bogged down. I drove 107 miles south on the Parks Highway before stopping and parking at the entrance to the Stampede Trail. We unloaded my cross-country tractor and began our wolf hunt as I drove the twenty miles to the Savage River.

We crossed the thirty-foot wide, one-foot-deep Savage River with no problem. We stopped in alarm at the snow and ice-coated banks of the wider and swollen Teklanika River, which had reached flood stage. The glacial silt gave a milky effect to the muddy river. "I don't think we can make it," I said to Bob.

With boyish enthusiasm he replied, "Sure you can." So we walked upriver, searching for the shallowest channel. Of course we really couldn't be sure because we could not see the bottom. We came to a wide place where the river spread out before dividing into two channels.

The crossing would be extremely risky. We couldn't afford to get the tractor stuck because it would spoil our hunt and I could lose my tractor and maybe our lives. We had absolutely no way of extracting the tractor from a swift running river. I expected the water to be no more than three feet deep and figured that's all the old Farmall tractor could take before the fan threw water over the distributor cap and the engine drowned.

Knowing that the river would be rising as the warm day melted snow up river, we decided to go for it. After elevating sleeping bags, clothing, and supplies to the highest point possible, I edged into the water. Almost at once the front wheels dropped down out of sight. I was in low gear so that I could stop and back out if the front end sank any deeper. Forty feet out in midstream, the water touched the bottom of the radiator. At sixty feet out I thought the water was shallower by a few inches. I was within twenty feet of the far shore and safety where I thought the water would continue to be shallower, and I pushed the gas lever a bit. The tractor picked up speed just as the front end dropped into a four-foot-deep channel. The water was halfway up the radiator. The fan threw a heavy spray of water even back to the driver's seat.

The distributor cap and the generator must have had a dirt and grease seal because they never damped out. The old alternator kept right on producing as we climbed out of the water and up the bank.

The stress melted from my face and I smiled at Bob when he blurted out, "See, I told you we could make it."

182

While silently recalling the fact that there was no other way out, I responded, "The Lord was with us this time, but I sure don't look forward to recrossing on our return trip." We rechecked our gear and emptied water out of pots and pans.

The Stampede Trail, built five years before, had never been kept up and some culverts had washed out. Boulders had rolled down every bank into the trail, which had been utterly neglected. But this was to our advantage because it became a thoroughfare, used extensively by the wolves, bears, moose, and caribou. All had left their droppings. I put the tractor in second gear and motored ahead at half throttle. I scrutinized the ground and noted the wolf scats. By the time we reached the old abandoned bus,* fifteen miles away we had seen three small bands of caribou and 150 or more piles of wolf scat.

Anticipation and excitement ran high when we found twenty-five scats within two hundred feet of the bus. From the bus I could see a beaver's house, and on the trail to it I counted four scats, two of them fresh. The house had been abandoned; not a beaver around.

We walked up the Sushana, a smaller creek, and spotted fresh wolf tracks and a fresh grizzly bear track on the sand. It had been a long day and we decided to make camp in the bus and cook supper while drying our wet clothing.

---

* Chris McCandless's decomposed body was found in this same bus by a moose hunter in 1992. Jon Krakauer assembled a national best seller book of Chris' life and times. Into the Wild is about a young man who hitchhiked to Alaska and walked alone into the wilderness and died in a deserted bus on the banks of the Sushana River.

∽ Exposing Alaska ∽

Bob was a fine camper and he volunteered to fry potatoes and onions since that is mostly what I brought to eat. They're one of my favorites, so for the next four days, potatoes and onions became the main course at each meal.

The old bus was a good dry camp. The old stove had a cracked top, which smoked, but still, it was wonderful to have a stove. There were also some clothes hangers. I cut some wood for the next day while Bob did dishes.

We then turned in, and I had the luxury of throwing my sleeping bag onto a well-worn set of Simon springs instead of the hard floor. The swishing of the river and an old owl hoot was about all we heard until morning, when we were aware of noises coming from under the hood of the bus. It was a mama porcupine talking to herself as she tried to make a nest where the motor should have been. We snapped her picture, ate breakfast, and started off up river.

We were following the wolf and bear tracks which both led upstream. About two miles up I saw a big hole in the right bank. I went over and saw the leg bones of a caribou. It was a secondary wolf den. The bone and the network of wolf trails leading away from it told me that it had been used previously. My heart pounded with excitement. The primary den I hoped would be within a five-mile radius.

We studied each and every sandbar and both riverbanks for the den that could be almost anywhere. A little further on, a grizzly sow and her cub came into the valley. That made three bears and a wolf (possibly a den) ahead of us. Our movements had to be cautious and calculating. I did not fear the wolves, but the grizzlies might ambush us. Further up the river we found the tracks of a fourth grizzly coming downstream.

184

∾ *Exposing Alaska* ∾

The wolf's tracks came and went as it scrutinized the potholes and crawled up the sandbars. The high water had washed out the concentration of old tracks on the sandbars.

At about two o'clock we turned around and came back to the bus camp. We had cruised all the rises and hills for eight miles up river and found no concentration of scats, tracks, or bones on the trails. Both of us grew wary of hunting in the brush-filled river bottom with so many grizzlies in the valley.

The next day we hunted downstream and only found one wolf track that led up a tributary but no concentration of wolf signs. We saw one grizzly track that day but could not catch up to the grizzly.

The third day we hunted on toward Stampede, some thirty miles away. We crossed two grizzly bear tracks and two wolf tracks, but when we climbed the hill the bears and the wolves had long gone.

The next day, our fifth day away from home, we started back. Two miles from the bus we found fresh wolf tracks and scat in the track my tractor had make four days before. Had we passed the den? A quick search of the creek upstream and downstream convinced us it was just old daddy wolf out hunting. We saw three sets of grizzly bear tracks and another fresh wolf track along the Teklanika but that wolf den appeared deserted.

The recrossing of the Teklanika River was just about as hazardous as before. The water had dropped six to eight inches and I still could not see bottom, but we crossed without a major incident.

We were trying to reach my truck in one day so we continued on, crossing the Savage River and heading for the highway twenty miles away. When we neared my truck we

185

got the surprise of our lives. There was Mr. Taylor's truck parked behind mine. He and some other men were out beating the bush in the canyon looking for us. He had the impression that we would be back in three days, so on the fifth day he had reported us lost and came looking for us. I got a lot of ribbing around town for the next few days and had my pictures printed in the newspaper.

About 7,500 wolves live in Alaska. They range in color from black to nearly white, with every shade of gray, tan and even "blue" between these extremes. Most adult male wolves weigh from 85 to 115 pounds, but occasionally reach 145 pounds. Wolves usually live in packs of about six or seven animals; however, a pack may have thirty or more wolves.

Caribou are stout members of the deer family, with concave hoofs that spread widely to support the animal in snow and soft tundra. The weight of adult bulls averages 375 pounds; females average 200 pounds. Biologists estimate that there are close to a million caribou in Alaska.

In the spring, the females migrate to the calving grounds. In late May or early June, a single calf is born; twins are very rare. Newborn calves weigh about fifteen pounds and usually double their weight in twenty days. They can walk within an hour of birth and after a few days can outrun a person and swim across lakes and rivers. Like most herd animals, caribou must keep moving to find adequate food.

Wolves and Grizzly

Paul Elbert

I was telling my neighbors Colbert and Ruth about all the grizzly bear tracks that we had seen. They wanted to get in on the fun, so we planned another trip for the coming weekend. We made it all the way to the bus by about midnight that Friday as it started to rain. We saw seven bands of caribou along the way in, a few new scats near the bus, and one small bear track.

It rained by the bucketful for twenty-eight hours, a steady warm rain. Numerous small glaciers in the foothills and ice along the shaded riverbanks melted swiftly. The volume of water that rushed down the valley was terrifying. By morning the Sushana was up four feet, chasing bears and wolves and moose out of the brushy areas along the river.

When the rain stopped the sun came out. We took off tundra-hopping over the ridges and hills. Our main quest this trip was the grizzly bears, but we did not sight one bear.

We did, however, see two small bands of caribou, five big bull moose, and three cows that day. One of the four big bull moose was running hard and when we got to the river bar we saw why. I counted five in the wolf pack that followed him.

Exposing Alaska

The next day we had to terminate the hunt except for hunting the ridge between Sushana and the Teklanika. About four miles from camp we saw a gray wolf running parallel to us, a third of a mile to the north. However, my companions needed to be back to work so we headed for the truck. We crossed the Teklanika with the water again over the distributor cap and the generator, and the fan gave us a shower. Four miles further on, we came to a peat bog where the road was washed out. We couldn't go around and there was no way to get through it.

Our only hope and salvation was to build a bridge, and this we were totally unprepared to do. Our one axe had hit a few too many rocks. However, we did find an old cross-cut saw in a dilapidated trapper cabin on the Savage River. It was rusty but sharp and with it we cut two of the biggest tress along the river. They were forty feet long and six or seven inches thick at the small end.

With the tractor I dragged them up to the washout. I got astride one and tried to push it up onto the far bank, but it kept getting fouled up in the mud and the brush underwater. We then went back to the trapper's cabin, found an old piece of plywood, took it to the scene, and by pulling the logs back a little and placing the plywood under the end, we directed the log up and into place.

Then we cut six key cross-logs and placed them four feet apart. We had not a single bridge spike or drift pin, but we did find a tangled mess of heavy wire behind the trapper's cabin and with it tied the cross-logs fast. Between these cross logs we then placed smaller cross-logs, which completed our bridge.

It took us a day and a half to build the bridge but it held the tractor and us and we were happy about getting back to

188

civilization. The road was washed out in several more places, but we were able to get through or around without major trouble or delay. This ended our second trip and we had not bagged a single grizzly bear. The caribou had begun to leave; however, sighting that gray wolf in the odd corner of the valley had set me afire with hope. I immediately began preparation for the third and final hunt.

The frail lady tourist asked Sourdough Sam, "How do you endure the bitter cold in Alaska?" Sam replied, "Madam, a person of good sense does not endure the cold, he protects himself from it."

We Found a Den of Wolf Pups

Paul Elbert

I selected fourteen-year-old Jack Mulley Jr. as my companion. I told him I had hunted half the valley out and had seen a wolf, one third of a mile north of the road. Jack, like his father, loved to hunt and trap and had inherited his dad's dislike for wolf predators.

Our trip in was uneventful. The caribou herds I usually saw along the way had all departed. We found a few fresh bear and wolf tracks along the road. We noted with amazement one wolf track with one pup companion in the mud where we had built the bridge. We found other lone wolf and lone pup tracks in the sand twenty miles further on. We were elated. Things were coming to a head. The next morning after an early breakfast we went out to track down the wolf den I had hoped would be one third of a mile to the north of the road, four miles back. In the morning we searched up one creek for scats, bones, tracks, and trails. In the afternoon we worked down the other creek. No signs, no den.

That night while we slept the big male wolf went upriver again, leaving his four-by-five-inch tracks on the sand. I spotted them in the morning and our hunt began in earnest. We worked up the river, slowly reexamining each sandbar for tracks and each stretch of woods for trails.

∽ *Exposing Alaska* ∽

About ten o'clock we found a network of old trails, indicating a wolf den had been nearby. After studying the area, I discovered what remained of a washed-out den in the riverbank, the casualty of recent flood stage water.

Now I was more confident than ever that we were closing in on a den. We were indeed closer than I had dared to guess. Proceeding up the gravel bars about two hundred yards I came to some sandy places covered with pup tracks. I looked over to the bank right into the yawning mouth of another secondary den. I walked over to it and saw it was full of porcupine dung and quills. I figured I must be wrong. Then I saw a flash of white or at least a very light color far back in the den. This I mistook for a rabbit. But a second later I saw two large front legs and thought this must be the mother wolf.

I had my 30.06 with me so I stepped back a ways so as to get a shot at whatever came out. About this time Jack returned after scrutinizing a hill. I pointed to the tracks in the sand and then to the den. The smile that spread over his face was a joy to see.

After eating a frugal lunch I left Jack to watch the den and shoot the mother or father if and when they returned. I took off upstream on a four-hour tour, hunting for the primary den. Way up stream I found the tracks of a lone wolf and a lone pup. But nowhere did I find the signs that indicated the location of the primary den.

When I returned at about 11:00 PM (We had just passed the longest day of the year, June 22), Jack had shot a hundred-foot roll of sixteen-millimeter movie film. He said that four very large pups had come out at different times to do their little jobs, look, and play around. The mother or

191

father had come close enough for a good shot but his gun had misfired.

We ate a midnight snack and took turns sleeping. It had started to drizzle and the small piece of Visqueen wasn't enough protection. When morning came we shot a few pictures as best we could, for it had now begun to rain in earnest.

At 9:00 AM I started to dig out the pups. First I had Jack push a ten-foot pole with two gunny sacks on the end into the hole as far as possible to plug it up. Then I estimated where the plug was and dug. My second hole was right over the pups. I reached down cautiously, expecting to be bit by the big pups. I got hold of hair right in the middle of one's back; then I slid my other hand up to the nape of the neck and pulled. Out came a twenty-pound pup, light gray in color and water blue eyes. It was a female. She did not attempt to bite or even squirm loose. She looked awfully frightened. I held her for a petting, which she tolerated. Jack snapped a picture then I dropped her in a gunnysack.

I reached down three more times and brought out three more pups. We now had the four Jack had seen outside the den. Before leaving, however, I checked once more and far back I saw feet once again. I reached in and put my hand gently on it and began to pull. There was a little protest, a little growl. But the leg came and soon I was able to get my other hand on the neck and pull. Out came the fifth twenty-pound, light gray wolf pup.

It was now raining hard and it would take us two long hours to return to the bus. When we got there the wolf pups were panting feverishly and one had chewed a hole in the sack and was about to escape. We transferred the pups to gasoline cases (wooden boxes large enough to hold two

Exposing Alaska

five-gallon cans). The pups were so large that their heads had to be turned sideways. We then nailed cleats across the tops. The pups were now very secure and very cramped.

It rained for sixteen hours, and the glacial Sushana River went on a rampage. We stayed in the bus for two days before attempting the trip home. We shot two porcupines and skinned them to feed the pups, but they did not eat or drink their water while we watched. If we left for an hour or so, they would eat and drink a little.

On the third morning we started home. But the Teklanika River was much too high to cross. So we camped under the cache next to a tumbled-in trapper's cabin. Behind the cabin was an old storehouse but the small roof poles under the sod roof had rotted and all had fallen onto the floor. I studied this for some time before deciding that we could turn the wolf pups loose in the ten-by-twelve-foot enclosure for a day. They were just too cramped in the twenty-inch long gasoline cases.

I shot another porcupine, skinned it and threw it into a far corner of the storehouse. Then one by one we opened the cases and thrust the pups through the three-foot-high window opening. All went well and we nailed the window shut. The pups fearfully huddled in a corner but displayed no eagerness to either climb or dig. After watching them awhile we believed they were secured.

Relieved, we zeroed in on feeding ourselves and since provisions were running out we fried the two hind legs of a porcupine for ourselves. Boy, were they tough to chew on. I nearly broke a tooth. At eleven o'clock I again looked at the pups. They had eaten their share of the porcupine but two of the pups were eying the top logs of the enclosure as if trying to conjure up a method of scaling them. I awoke at

193

4:00 AM and looked again. The pups were peacefully sleeping, and I crawled back into my sleeping bag feeling secure and very happy.

When Jack awoke at about 7:30 and looked, the pups were gone. While violently shaking me awake, he hollered, "They're gone, Paul; get up, they're gone." Jumping up in an alarmed state I banged my head into the bottom log of the cache. In dismay and disbelief upon reexamining the makeshift enclosure I spotted a hole right through the bottom log. It was rotten and soft as mud under the hard bark, which had given us the illusion of sturdiness. Once the pups tore through the bark they easily clawed right out. We searched all around but the ground was covered with moss and grass, and it was impossible to even tell which way the pups had fled.

Two very disappointed wolf hunters began an FBI-style investigation. We searched the sandbars along the river. We walked the road and the trails and began to lose any hope of ever seeing or catching them again. At 7:30 in the evening we walked a mile down to the river to see if the flood was receding. The boiling water was up a couple of inches. Dejectedly we turned and walked back. After proceeding some two hundred yards we were electrified by a wolf pup wail: a long, drawn-out pitiful cry for its mother.

Jack looked hopefully at me and in a whisper asked, "Paul, is there any way to catch them again?" I said, "Circle far around and come in back of them. I will approach from here. They will dive into a hole or pile of brush the minute they become aware of us."

Jack circled even farther than I had anticipated but it proved to be exactly right. Just as his eyes came over a little rise on the hillside he saw a pup on the next rise two hun-

194

∾ *Exposing Alaska* ∾

dred feet away pointing its nose to the sky and wailing. Then he saw two of the pups streak into an old porcupine den. It was a big den that spread under three successive trees along the edge of a five-foot bank.

Jack ran over to the den and poked brush and moss in the nine or ten different entrances. He yelled and yelled for me. I could not understand his words, but I reasoned he wouldn't holler unless he had something bottled up. I got to the top of the hill out of breath. Jack said, "I saw two of them run towards this den. I didn't actually see them enter. They may have gone by."

I put my ear to the ground and listened. I could hear something breathing and scratching down there. I guarded the holes so that nothing could escape and studied the surrounding area. I saw that fifty or more spruce trees had been barked, some recently. I didn't know if our digging would produce wolf pups or a family of porcupines.

Jack fetched the ax and shovel in jig time and we began digging. Two and a half hours later, after digging all around two of the spruce trees, we finally cornered the animals under the third tree. Fortunately the animals turned out to be wolf pups. The second digging out of these pups was ten times harder than the first time, but we succeeded. I lifted the pups out one by one as before and dropped each into a gunnysack and tied it securely. We only found three. We went back to the other trees to see if there was some off chamber that I had missed, but no, two pups had disappeared. It was now almost midnight with still plenty of light since we were close to the Arctic Circle in the land of the midnight sun. We carried our three pups back to camp, put them back into gasoline cases, and nailed two cleats across each.

195

We felt mighty happy having three of our wolf pups back. Before going to sleep that night I began evolving a method of taking some pictures by reenacting the escape. In the morning I made a pen out of old tin roofing around the backside of the storehouse the wolves had escaped from. I set up my 16 mm Bolen movie camera and focused it on the hole. Jack turned the three loose in the building again, and in just seconds they came wiggling out, one after the other. I shot a whole hundred-foot roll and although the light on the north side of the house was not the best, the pictures are quite good.

Later in the day I again climbed the hill where we had last dug the pups out. I was looking for some clue as to what had happened to the two females that had disappeared. Four or five hundred feet further on I found a mudslide and the pup tracks with those of a large wolf, which was evidently leading them off to safety.

Antlers Coming

Paul Elbert

I straightened up and felt a crushing pain in my chest from the strain of packing out the last hindquarter of moose the previous year. My first heart attack. It was not too severe, but from that day on my physical exertions were of necessity, cut in half. In any future efforts to engage in the thrill of hunting, I was now forced to become a still hunter.

So, early this September high in the hills where timber feathers out into brush, where the nearly snow-covered hills are mirrored in small sapphire lakes, where the game trails converge in the very heart of this big plateau, I built a hunting blind and waited.

I have dubbed this place the Trophy Pasture because grizzly bears and hungry wolves have eliminated the weak and the young. The Dall sheep and caribou are almost annihilated. Only the big, strong, and hearty animals remain. My heart gets to racing with excitement when I look out across the expanses of brush and see moose antlers coming. The spread is apt to measure sixty or more inches. These antlers have stopped growing because the velvet has dried up at their base and has begun to peel off in half-inch ribbons. The rut (mating season) is near and these big bull moose strut up and down the valley, eager to fight and prove their prowess and their right to the favor of as many moose cows as they can convince.

❧ Exposing Alaska ❧

Overflowing with anticipation, I hoped to lure a trophy right up to my blind by flailing the brush with a club and grunting to simulate the cow moose call. Twenty-two years of successful hunting in Alaska have not lessened the thrill that surges through me from watching white antlers coming from a mile away; nor has my enjoyment and the feeling of accomplishment I get while butchering out a prime bull moose for my winter's meat decreased. I still photograph and measure the antlers.

Fresh large moose tracks, pressed deep into the game trails, speak well for the likelihood and success of my hunt. Winter had stepped down from Mt. McKinley, letting her white skirt fall onto the nearby foothills. Icicles framed the waterfall and a gentle mountain breeze whispered of snowflakes.

The high altitude and light air released a weight from my body so that my old bilge pump sent a surge of red blood to stimulate every cell. To me this was ecstasy and I was enjoying the pure excitement of being fully alive. The sun scooted down like a golden spaceship making a slow landing. After that, a gold wash lingered above the western hills and sporadically, large flocks of geese and cranes swished south. Eventually the distant stars came down and the aurora borealis slithered out to sweep the sky with a broom of fire. I watched in wonderment as the day ended.

The next day, my body needed more rest so I took a two-hour-long siesta during an afternoon shower. When I awoke, the sun had made the freshly washed woods and tundra glisten. I resigned myself to try to decipher the line of communication the Creator had used. I suddenly realized the beautiful panorama I feasted my eyes upon was but the back of the Creator's canvas. I can only conjecture about the beauty of details to be seen on the Creator's side.

The light changed and I became aware of the Creator's silhouette around and through the canvas of creation.

Long hours and even days passed before the trophy ambled by, so I had time to resolve my inspirations into a concise philosophy. Inspiration comes by the thimbleful and I wrestled with it to discern whether it be wisdom or foolishness. The self-evident purpose of life is to enjoy it, to pursue happiness, a state of well being physically, mentally, morally, domestically, socially and politically. I see but the back of the Creator's canvas. His silhouette shines through every living thing. He repairs and maintains every cell of my body as long as I live.

I awoke from my trance, aware that antlers were coming. The moose was white: an albino. This I took as a tangible sign that the Creator approved of my meditations.

～ Exposing Alaska ～

Paul Elbert at his cabin on the homestead with a set of moose antlers and a black bear hide.

In the fall of the year, the geese gather and head south to warmer weather.... You also notice the snow birds flocking in the fall. They are the wiser Alaskans who can afford to head south for the winter. Sourdoughs scoff at snowbirds, until they can become one. The state motto is "North to the Future." The snowbird's motto is "south for the winter!"

The Author

Don Elbert moved from Seattle to Fairbanks, Alaska, in 1947 when he was thirteen years old. He came to help his dad farm a hundred-acre homestead. Don learned how to work, clearing land, growing potatoes, cutting firewood, hauling water from a creek bed, and surviving in a cold climate.

For the past fifty-five years, Don has experienced the essence of Alaska by climbing its mountains and walking through its valleys. Don retired after thirty-one years as a heavy equipment mechanic and still lives on one acre of the original homestead near Fairbanks with his wife Elaine. They raised five children and now have sixteen grandchildren. Over the years Don has farmed, panned for gold, boated, fished Alaska rivers, hunted and trapped game, and repaired a lot of equipment here in the north.

Don learned to love this land, like his father did before him. This book is his way to share with you some of the best and the worst of the adventures and spectacles that for him made life worth living in Alaska.

Other books published by Don Elbert:
    Fascinating History, Facts, and
        Fun About Alaska—in 1995
    Alaska: The Cold Truth—in 1996
    Fascinating Alaska: Inhale the Essence—in 1999

～ Exposing Alaska ～